Paper-based Diagnostics

Kevin J. Land
Editor

Paper-based Diagnostics

Current Status and Future Applications

Springer

Editor
Kevin J. Land
Council for Scientific and Industrial Research
Pretoria, South Africa

ISBN 978-3-319-96868-1 ISBN 978-3-319-96870-4 (eBook)
https://doi.org/10.1007/978-3-319-96870-4

Library of Congress Control Number: 2018959868

This Springer imprint is published by the registered company Springer Nature Switzerland AG
The registered company address is: Gewerbestrasse 11, 6330 Cham, Switzerland

Preface and Acknowledgements

The paper and pulp making process was first developed thousands of years ago, and through the centuries has spread throughout the world. The manufacturing processes have been optimised, as have the materials which are used. Paper is today used in hundreds of applications, ranging from the printing of money, to packaging, as a storage medium in the form of writing material, to cleaning materials such as toilet paper and paper towels.

Paper has also been used for performing various diagnostic tests, with the litmus pH test and pregnancy lateral flow tests being best known.

Recently, the possibility of developing paper-based diagnostics has received attention due to the many benefits of paper, such as its low cost and recyclability.

The World Health Organisation (WHO) ASSURED (Affordable, Sensitive, Specific, User-friendly, Rapid, Equipment-free, Delivered to end users) criteria have also led to many discussions as to how to achieve diagnostics which can be deployed in resource-limited settings. Paper is often mentioned as a substrate material of choice. However, there are still many developments required in order to successfully develop tests which can be deployed where they are most needed.

There have been a number of excellent reviews published since the Whitesides group first reintroduced micro paper analytical devices in 2007. However, there is little in the way of complete resources. The motivation behind this book is to start developing a reference which future researchers and developers can use in their efforts to perfect paper-based diagnostic tests. It should serve as a summary for people wanting to get into the field, while also serving as a valuable resource for established researchers. It is clear that this remains an active and early research field, and that the real-world benefits which can be gained from developing these tests will indeed be life changing and will rapidly begin emerging into products.

The first chapter "*Unmet Diagnostics Needs for the Developing World*" describes the need for paper-based diagnostics with a focus on health and the developing world. This chapter shows an in-depth view of the field while giving a brief history on how paper tests have evolved. The chapter also unpacks the WHO ASSURED criteria in detail. The authors use the SWOT methodology to describe the Paper Diagnostics, so that the reader is aware of the weaknesses and threats, together with strengths and opportunities.

In Chap. 2, "*The Role of Paper Chemistry and Paper Manufacture in the Design of Paper-Based Diagnostics,*" paper manufacturing and paper chemistry is described in detail. The reader will gain a background understanding of the processes used to manufacture paper and also insights into how paper can be engineered in order to add functionality to make it more useful for paper-based diagnostics.

In Chap. 3, "*Critical Components and Innovations in Paper-Based Analytical Devices,*" the authors describe the key innovations and advances which have allowed paper to move from being used in niche applications in laboratories to where it is a powerful tool used in many different disciplines and many different applications across the world.

Chapter 4, "*How Are These Devices Manufactured?,*" takes the reader through the manufacturing methods which have been developed to manufacture paper devices. Additionally, the main characteristics of each method are described, and costs associated with different methods are detailed. This is critical as cost will eventually be one of the deciding factors for determining the successful market uptake of these diagnostics.

In recent years there have been rapid developments in printed technology such as: printed electronics, communications, industry 4.0, and the internet of things (IoT). Chapter 5, "*Printed Functionalities on Paper Substrates Towards Fulfilment of the ASSURED Criteria,*" describes these developments in detail and suggests ways in which they can be utilised to add additional functionality to paper-based devices.

In Chap. 6, "*The Many Roads to the Ideal Paper Device(s),*" the authors investigate recent developments in sample preparation, sensitivity and specificity, and describe ways to overcome currently experienced problems in order to arrive at the ideal device.

Finally, in Chap. 7, "*The Fascination of Paper,*" paper is looked at not from a strictly scientific perspective, but from an alternative perspective which explores why we still find paper so fascinating, even though it is so part of our everyday lives.

There is some overlap between chapters, especially with regard to future challenges and the way to overcome these. This is intentional, as each chapter describes different aspects and tools available for the researcher hoping to overcome the issues, and it is hoped the reader will thus be in a position to approach solutions from different, fresh perspectives.

In this respect, the book aims to serve as a catalyst to solve the remaining challenges and position paper-based diagnostic devices to become powerful tools in the hands of health care (and other sector) professionals where they will make significant impact.

As always, there are a great many people who need to be thanked.

Chuck Henry and Barry Lutz for insightful discussions during the early stages of the book.

The Council for Scientific and Industrial Research (CSIR), South Africa, for permission to take on this task.

All the authors who have enthusiastically offered their time to write their chapters. This is no easy task given modern day-to-day pressures.

The staff at Springer who have been supportive throughout the process.

In closing, it would be fantastic if this book inspired people to get involved in this exciting field, and that devices based on paper could be produced which would have real influence, especially in resource-limited environments where they can have the most impact. It is also ideal that the book would be read by people not directly involved with the field, such as policy makers and funders, and that the awareness created on the importance of the work could encourage major new focus and funding streams to be made available for development. There is much room for debate.

Contents

1 **Unmet Diagnostics Needs for the Developing World** 1
Kevin J. Land, Suzanne Smith, and Rosanna W. Peeling

2 **The Role of Paper Chemistry and Paper Manufacture in the Design of Paper-Based Diagnostics** 23
Samuel Schabel and Markus Biesalski

3 **Critical Components and Innovations in Paper-Based Analytical Devices** 47
Robert B. Channon, Monpichar Srisa-Art, Katherine Boehle, and Charles Henry

4 **How Are These Devices Manufactured?** 89
Giorgio Gianini Morbioli, Thiago Mazzu-Nascimento, Amanda M. Stockton, and Emanuel Carrilho

5 **Printed Functionalities on Paper Substrates Towards Fulfilment of the ASSURED Criteria** 123
Suzanne Smith, Dario Mager, Jan G. Korvink, and Kevin J. Land

6 **The Many Roads to an Ideal Paper-based Device** 171
Margot Karlikow and Keith Pardee

7 **The Fascination of Paper** 203
Jan G. Korvink

Index 215

Chapter 1
Unmet Diagnostics Needs for the Developing World

Kevin J. Land, Suzanne Smith, and Rosanna W. Peeling

1.1 Introduction

It has long been recognised that diagnosis is a critical component of healthcare, and without sufficiently accurate diagnosis treatment is difficult or not possible. Advanced laboratory-based systems have been developed over many years and now for many different analysis requirements and analytes it is possible to get a quantitative analysis. Development has progressed very rapidly, but equally rapid has been the increase in cost and, in many cases, the complexity of the tests. As a result, diagnostic tools with high accuracy and precision are commercially available to guide treatment and management decisions for many diseases and conditions, but they are neither accessible nor available to patients in resource-limited settings.

Diagnostics are traditionally undervalued both in the developed and developing world. A survey conducted by the Lewin Group showed that although diagnostic results are used in 60–70% of all clinical decisions in the USA, the spending on diagnostics was less than 5% of hospital costs and less than 1.6% of all medical care costs [1]. The results obtained from these diagnostics have far-reaching benefits, not just to clinicians and patients, but to healthcare managers and disease control and prevention programmes in general. The benefit of screening for diseases early before complications develop or to break the chain of transmission was confirmed in a later report by the same group [2]. In contrast to a diagnostic test for patients presenting with symptoms, a screening test is performed within a population to identify infections before symptoms become apparent. A good example is the use

K. J. Land · S. Smith
Council for Scientific and Industrial Research, Pretoria, South Africa

R. W. Peeling (✉)
London School of Hygiene and Tropical Medicine, London, UK
e-mail: rosanna.peeling@lshtm.ac.uk

K. J. Land (ed.), *Paper-based Diagnostics*,
https://doi.org/10.1007/978-3-319-96870-4_1

of simple rapid tests to screen for HIV so that patients can be referred for treatment and disease transmission to others in the population can be prevented. Such tests are critical in the prevention and control of the HIV epidemic. More recent work confirmed that in vitro diagnostics accounted for 2.3% and 1.4% of healthcare costs in the USA and Germany, respectively, while 66% of decision-making was based on these results [3]. This is a surprisingly low percentage of healthcare costs given the benefit of the diagnostic.

In 1995, the Rockefeller Prize was launched in order to develop a test for asymptomatic chlamydia and gonorrhoea [4]. In 2003, the ASSURED (*A*ffordable, *S*ensitive, *S*pecific, *U*ser-friendly, *R*apid and Robust, *E*quipment-free and *D*eliverable to end users) criteria was published by the World Health Organisation (WHO) as a guideline for the development of tests for the developing world [5]. This ASSURED criteria has been used by many research programmes in developing their own tests, although often not meeting all the requirements. For example, more sophisticated point-of-care devices may be highly sensitive and specific, but may be higher in cost, while standard lateral flow tests are low in cost but may have limitations in terms of accuracy and result interpretation. This is explored further in this chapter under the WHO ASSURED section. Additionally, the criteria have been used for the development of tests for other diseases and in other areas (such as environmental testing) and have often been used as a foundation on which to develop diagnostics for resource-limited settings [6]. More recently, important requirements have been identified in addition to the existing ASSURED criteria to take into account significant technological changes over the last 10–15 years. These aspects include device connectivity and environmental friendliness, as well as the ease of specimen collection for testing. Scale-up of devices for mass production and distribution also needs to be considered.

Diagnostics which can be performed at the point of care outside of laboratory settings have made a significant contribution to increasing access to diagnostics but are still costly in terms of test manufacturing as well as transportation and storage. In addition, they are often not environmentally sustainable. Paper-based diagnostics with advances in printing technologies will provide alternatives to the current generation of point-of-care tests.

In this chapter, we discuss the critical role of paper in the future development of these tests, and why it could be seen as the substrate of choice for the development of future biosensors. Paper is abundantly available, of low cost, easily disposed of, and easily modified for enhanced flow and detection and manufacture can be scaled to produce billions of devices per year. Current research and development indicates that power sources such as batteries, screens for visually displaying results, electronic circuitry, biological materials, flow-directing structures, and many other functional attributes can be added to paper simply utilising printing technologies and other manufacturing techniques which are readily available and well understood.

Paper-based diagnostics will play an important role in the eventual goal of solving many of the world's greatest health challenges, which are directly connected to resource-poor environments and the need to measure and diagnose people and the surrounding environment. We discuss the different settings in which this technology can play an important role, the role which paper can play, together with advantages and disadvantages of paper, and where the future technology development challenges will be.

Paper-based diagnostics will not replace all point-of-care (POC) diagnostics, and should rather be thought of as a technology which can add enhanced functionality and access where conventional tests struggle to meet requirements. It has great potential for application to some key diagnostic tests. This chapter focuses on examples that illustrate how paper-based diagnostics could make an impact in resource-limited settings.

1.2 WHO ASSURED Criteria: Success to Date and Remaining Concerns

Paper-based diagnostics have a number of attributes that are well suited to fulfilling the ASSURED principles. ASSURED was developed as part of the WHO Sexually Transmitted Disease Diagnostics Initiative, but is typically utilised for any test to be used in resource-limited settings [7]. The ASSURED criteria are summarised in Fig. 1.1. These criteria have been accepted as a benchmark to which the ideal test can be compared.

Figure 1.2 describes each one of the ASSURED criteria, together with colour bars to indicate what is achievable in a particular healthcare setting (urban, semi-urban, or rural). Ideally all these criteria are important for all three settings. Since no test is perfect, and there are constraints at each setting, trade-offs need to be considered. In a rural setting, tests need to be rapid as patients often travel long distances to get to a health facility and cannot easily return for a follow-up appointment to get test results and treatment; heat stable and equipment free as there is often no electricity to operate equipment or refrigeration; easy to perform as there are often no skilled technicians; and affordable as rural populations are often not as financially stable as urban populations. Under these constraints, it is often not possible to have tests with high accuracy. In urban settings, there are laboratories with skilled technicians to run highly accurate tests on sophisticated equipment, so tests with high sensitivity and specificity can be available even if they are more expensive.

Although largely accepted, there have been criticisms of the ASSURED criteria. As an example, ASSURED only address the characteristics of the desired test but do not directly address sample preparation, skills required to operate the tests, or clinical decisions which need to be made.

In addition, no attempt is made to define how the various aspects of the criteria might be designed into a single test. As an example, equipment free and need to retain sensitivity and specificity are criteria which may conflict with each other, or make it very difficult to design and produce a test which then still remains

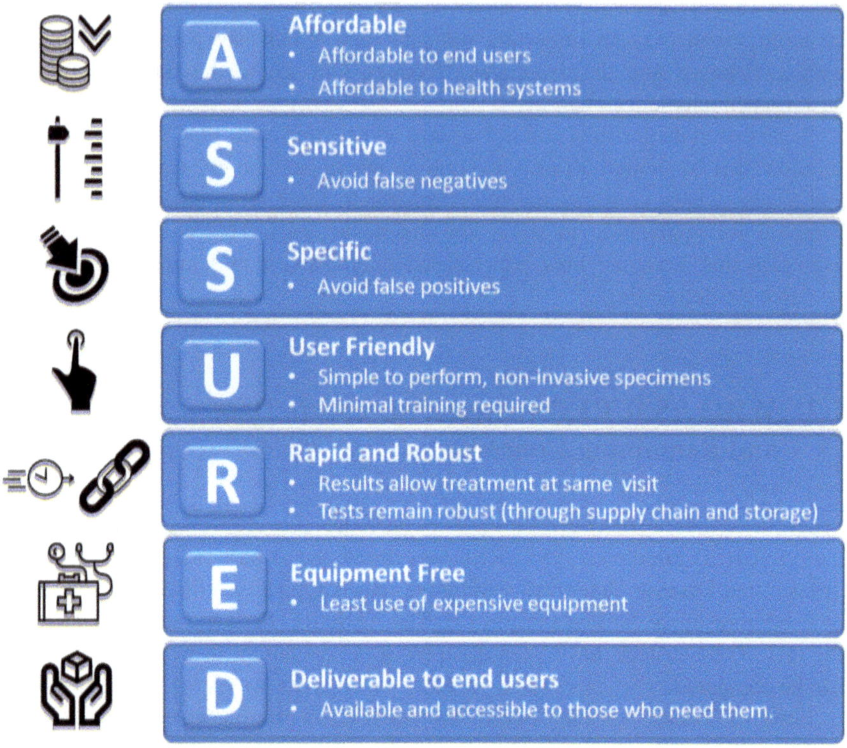

Fig. 1.1 The WHO ASSURED criteria to which all newly developed diagnostics for resource-limited settings should aspire to conform. Graphics: Noun Project: hunotika, Andrejs Kirma, Andrew Hainen, Michele Zamparo, Marie Van den Broeck, Bonegolem, Surangkana Jomjunyong, Arthur Shlain

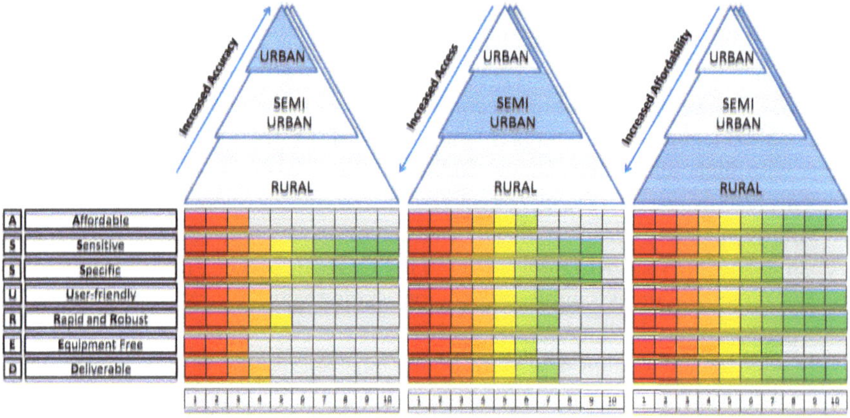

Fig. 1.2 The ASSURED criteria and their relative importance in different healthcare environments. Ideally these criteria are important in all settings. However, what is needed (affordable, rapid, heat stable, easy to perform, equipment free) and achievable (accuracy) varies by setting

affordable. So the criteria present a formidable set of challenges to any diagnostic designer wishing to adhere to all the requirements.

More recently, additional criteria have been introduced as an update to the original ASSURED criteria to take into account technological advances made in the last few decades. These include aspects such as environmental friendliness, real-time connectivity of devices, and scalability.

1.2.1 Environmental Concerns

Environmental friendliness of point-of-care diagnostic devices is an important consideration, particularly as usage of these devices increases. Paper devices have many inherent advantages, including environmentally friendly characteristics such as disposability and biodegradability. Currently, however, the majority of lateral flow strips are individually packaged within plastic cassette housings which cannot be recycled or incinerated as toxic fumes and substances may contaminate the groundwater. Additionally, these cassettes are then packaged into a foil pouch, which serves to keep humidity to a minimum as well as provide mechanical protection to the cassette and lateral flow strip. Besides this packaging, the test would typically contain a desiccant and possibly other reagents for use with the test. After use, all these components need to be disposed of safely. This can present a problem in resource-poor environments, and a large number of these components end up in landfills or are otherwise not safely disposed of, resulting in hazardous waste.

It is expected that the use of these types of tests will increase considerably in the future, further raising environmental concerns. Currently, over 100 million of each of HIV and malaria rapid tests are used every year, the vast majority being in resource-limited areas. Although these two are the most common tests performed, there are a great many other tests, all of which are packaged in plastic and aluminium foil.

It is suggested that more effort is devoted to developing environmentally friendly packaging materials, or reducing the volume of current material usage. Paper-based tests and paper-based packaging products would be ideal. Recently, considerable research and development effort has been conducted in the area of nanocellulose. It is renewable, biodegradable, and nontoxic, and also exhibits other desirable properties such as having a tunable surface chemistry [8]. Additionally, nanocellulose has shown potential for use in the area of printed electronics, both as a substrate material and as a functional material when blended with conductive particles [9]. Besides new materials such as nanocellulose, it is also possible to coat paper with biodegradable coatings, thereby allowing for replacement of synthetic polymer products [10]. Material usage can be minimised by combining several product elements into one single element. As an example, lateral flow strips could be packaged directly into sleeves which provide the functionality of the cassette, foil pouch, and desiccant.

1.2.2 Connectivity

Connectivity is an important consideration for development of future diagnostics to enable collection and transmission of test data for surveillance and outbreak alerts, as well as for capturing disease trends to assess the impact of disease control interventions. A secondary benefit of data transmission is the opportunity to build in alerts for stock management and to link in proficiency testing data to each site so that remedial training can be implemented if testing quality becomes suboptimal. This is especially important for testing sites where staff turnover is high. Bidirectional data flow would contribute to making devices more intelligent, also allowing for tracking of devices and potentially monitoring environmental conditions [11, 12].

1.3 The Promise and Versatility of Paper-Based Diagnostics Technologies

The drive towards paper-based diagnostics is not surprising given the many advantages that paper offers, particularly for resource-limited settings where these devices are most needed. In these settings, electricity ranges from nonexistent to intermittent to stable. Devices need to be able to handle frequent power surges and dips, including being able to come on line after power failures. Most equipment are not tolerant to this, and this adds to the challenge of maintenance and repair requirements for point-of-care instrumentation. This also requires additional skills and is time consuming, increasing the costs associated with the deployment of point-of-care diagnostics. Where it would be possible to replace equipment with single-use paper tests, many of these problems would be alleviated, as there would be no need for maintenance or repair.

Where power would be required to drive some of the device components, an alternative to highly robust external instrumentation would be to have power built into the device itself. Printed and paper-based batteries, biofuel cells, and solar cells have been explored [13, 14] and show potential for powering paper-based diagnostic devices where detection or readout may be required.

An additional advantage inherent to paper is that fluidic flow is achieved through capillary action and does not require pumps or external actuation. Not only does this enable automation of fluidic processing to a large extent, but also negates the need for electricity/power that would otherwise be required to propel and control fluids through the diagnostic device. This means that required on-board power for paper-based diagnostics would be lower than in other devices where fluidic actuation is required, and thus disposable, low-power, and low-cost power sources would typically be sufficient.

The development of various functional printable materials has also enabled paper devices with added and automated functionality to be realized. As an example, pressure-sensitive and temperature-sensitive inks have been developed, allowing sensors to be implemented directly on to paper devices. Positive temperature coefficient (PTC) inks enable heating elements to be printed on to paper, which is an important capability for many disease diagnostic test processes, such as polymerase chain reaction (PCR). Paper-based actuators including switches and valves have also been investigated, with the potential to automate processing or user interaction steps. This could potentially lead to automated sample preparation in future devices, reducing the skill level required to perform the test.

Paper can also be modified during manufacture or during post-processing to alter the physical and chemical properties of the paper, making it suitable to different reactions and reagents, as well as assisting in optimization of printing of fluidics and electronics on to the paper substrates.

Given these beneficial properties of paper, successful paper-based point-of-care diagnostic devices are likely to be realized. If this is to happen, a number of other challenges will need to be solved. This will include aspects such as quality control of the manufactured devices, along with regulatory reviews of safety, effectiveness, delivery, and service logistics. If single-use paper tests can be developed that require little to no external instrumentation, these challenges will be simplified to some extent. In addition, paper is lightweight and stackable, making many of the logistical considerations less complicated. The disposable and biodegradable properties of paper also simplify logistical considerations and provide a sustainable solution for future generations.

Acceptance by users of these new technologies and test formats along with the resulting increase in clinical work load will also need to be addressed. The increase in workload and time taken to perform tests using different point-of-care diagnostic devices for HIV testing in a South African clinic setting has been highlighted [15]. Kumar et al. presented two case studies of paper-based diagnostics tested in resource-limited settings, highlighting the various hurdles to be overcome for successful deployment and acceptance of these devices [16].

Functional paper-based components and systems towards point-of-care diagnostics are well under way, but the long-term deployment plan for these devices needs to be explored in the early stages to ensure that the developed technologies are effective and utilized to their full potential.

The various strengths and opportunities of paper-based diagnostics compared with other point-of-care diagnostics can be summarized (Table 1.1). In addition, the various challenges faced for successful paper-based diagnostics to be realized are also highlighted in the form of weaknesses and threats to provide a SWOT analysis of paper-based diagnostics. This analysis will serve as a good starting point for discussions around paper as a substrate for diagnostic systems.

Existing point-of-care tests can also be studied to determine where these tests comply with the stated requirements. Laboratory-based tests are typically expensive and equipment based, but provide sensitive and specific results. Lateral flow tests (such as pregnancy, malaria, or HIV lateral flow devices) are inexpensive, easy to use, deliverable to end users, and equipment free, but may not always provide sufficient quantitation in terms of results and can be prone to errors in result readout. So it would be useful to combine the best properties of tests from the different

Table 1.1 SWOT analysis for paper-based diagnostics

Strengths	**Opportunities**
Low cost	Easy distribution to remote sites
Usage not reliant on laboratory facilities	Closely linked to other emerging technologies (e.g., printed electronics)
Single-step sample preparation	Decentralised testing
Equipment free	Manufacturing on a distributed basis
Used by unskilled people	Maintenance free
Paper is easily available	Paper can be used for prototyping
Instrument-free fluidic transport	Underutilised as a testing platform
Biocompatible	Excellent tool for education
Biodegradable	Possibility of more than simple yes/no answers
High surface-to-volume ratio	Multiplexing of tests
Disposable	Manufacturing using conventional, well-understood processes
Conforms to WHO ASSURED criteria	Scalable
Weaknesses	**Threats**
Sample retention	Regulatory approvals
Not always ideal in wealthy first-world environments	More expensive, more sensitive and specific equipment (competitor products)
Technical hurdles to be overcome	Technical hurdles take longer than expected
Quantitative results not always possible	Batch-to-batch consistency
Difficult-to-build business model	Market (health services and end user) acceptance
Flow control	
Compatible barrier materials	

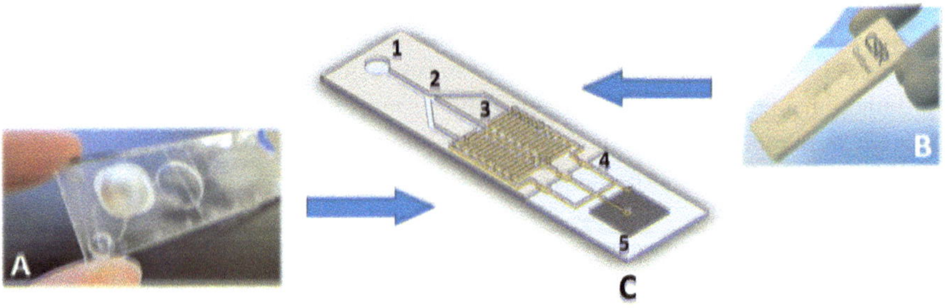

Fig. 1.3 Combining elements of different tests ((**a**) machine-operated microfluidic cartridge, (**b**) paper-based lateral flow) to create the ideal ASSURED test (**c**). As an example, the ideal test would incorporate (1) sample introduction, (2) printed microfluidic channels, (3) printed sensors, and (4) printed electronics, as well as (5) communication and connectivity options

environments into a test which conforms more closely to the full set of ASSURED criteria. Figure 1.3 represents this idea conceptually to advance the thinking around the construction of such a test.

Existing technologies need to be adapted for resource-limited use, opening up a huge opportunity to develop new technologies which can immediately address some of the challenges presented in adapting existing technologies to new environments. Many, if not most, of the technologies required to accurately diagnose and conform to ASSURED are available today but it is critical to integrate these technologies into operational devices.

Technologies including customization of paper substrates and the field of printed electronics provide a foundation on which to implement integrated ASSURED devices. Paper can be developed with favourable properties suited to different fluidic and functional requirements through printing and manufacturing processes, while printed electronics have advanced to enable functional components and complex integrated systems to be developed on to flexible and paper-based substrates. Electronic components, sensors, displays, lighting, power, and communication modules have all been demonstrated with some success [17], and can be leveraged to realize integrated paper-based devices. These functional printed and paper-based technologies are summarized at the end of this chapter (Fig. 1.6) to provide an overview of the various components that are envisaged to make up integrated ASSURED devices.

Given these emerging technologies that could be utilized to realize ASSURED devices, it is important to assess whether these technologies are accessible in resource-limited settings. Extensive investigation is being undertaken in South Africa and other countries to study the best implementation methods for point of care and the challenges currently faced [15, 18, 19].

Is there a way in which resource-limited settings can leapfrog technology hurdles, thereby achieving the same level of diagnostic intervention possible in far more developed countries? Given the constraints present in resource-limited settings, such as funding, infrastructure, and skill development, it is imperative to develop technologies which put a lesser burden on these requirements. Paper diagnostics would be one such intervention.

The combination of, for example, paper and printing technologies opens the door for massively scalable manufacturing, with roll-to-roll printing processes being an established technology. It is the merging of the technologies—for example the integration of paper-based sample processing and printed electronics—which will create the new paradigm, as the majority of the individual technologies are well known.

Typically, solutions are developed by the first world for resource-limited settings, with the first world retaining the resulting technologies. The ideal would be for countries with large resource-limited areas to manufacture and distribute their own technology solutions, and for paper-based technologies this could be feasible, as existing large-volume manufacturing techniques such as roll-to-roll screen printing can be utilized.

In addition to the importance of manufacturability and scaling, the deployability of the final product is crucial. In this respect, paper also offers many opportunities with ease of transport by stacking of lightweight devices as well as disposability.

1.3.1 Technologies Already Commercialized

Glucose testing has a long history, and is well described in literature [20]. To set the scene for paper-based diagnostic development, and showcase how development may progress, it is of interest to briefly describe the glucose test pathway. Figure 1.4 shows the development timeline. Of interest is not only the time it took to get to the present testing methods, but also the substantial development which continues in order to advance the next generation of glucose tests (e.g. continuous and non-invasive testing).

In a similar way, paper-based diagnostic systems have developed and this timeline is shown in Fig. 1.5. Significant milestones were the development of the dipstick test, lateral flow test, and more recently the micro paper analytical device or μPAD. Paper-based tests include well-known dipstick tests commonly used for pH testing of water or urine, as well as lateral flow strips utilized for pregnancy testing and rapid HIV testing. Thorough overviews of the design of lateral flow test strips, the manufacturing equipment utilized to produce these devices, as well automated instruments used to read out and quantify the lateral flow test strip results are available [21]. In addition, integrated lateral flow devices are explored, utilizing automated sample introduction and reagent storage mechanisms [22].

More recent paper-based devices or μPADs enable multidimensional fluidic flow and multiplexing through stacked or folded paper structures, also referred to as origami paper-based microfluidics. When compared to traditional lateral flow devices where flow moves laterally and tests can be difficult to multiplex, these devices facilitate the addition of functionality to the paper, either passively or actively. However, there are many overlaps and the two technologies are rapidly converging. Traditional lateral flow is adopting many ideas and functions from paper-based devices, and likewise paper-based devices are adopting many ideas and

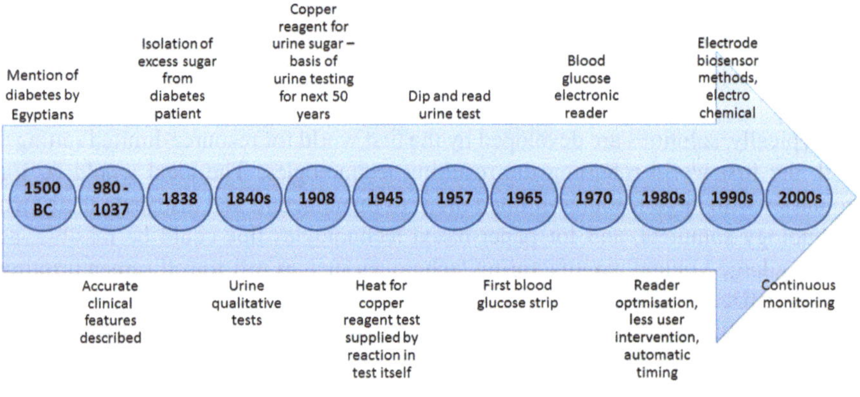

Fig. 1.4 Brief timeline of major developments in glucose testing

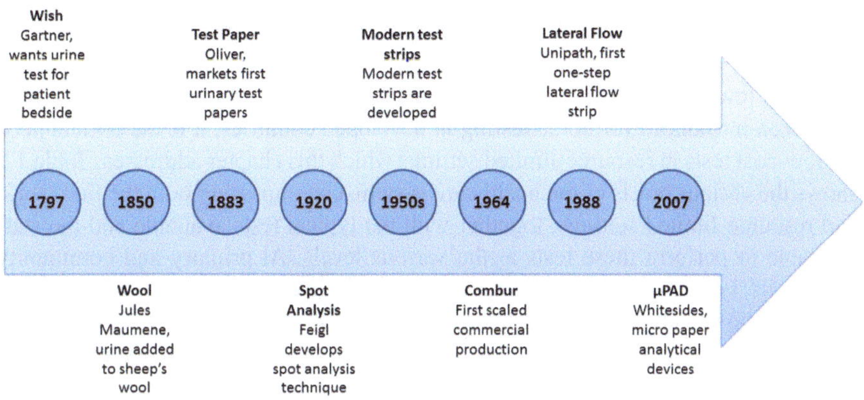

Fig. 1.5 Timeline of paper-based diagnostics development

functions from traditional lateral flow. Comprehensive reviews on the evolution of paper-based diagnostics are available, including recent overviews of paper-based diagnostics [6, 23].

1.3.2 What Is in the Pipeline?

The timelines shown in Figs. 1.4 and 1.5 affirm that these are not trivial development pathways and that realistic planning and projections need to be taken into account while being adaptable to future technology trends and unforeseen developments or innovations.

The development of smart paper-based sensors requires a roadmap or development pipeline to identify what is required in terms of formats, functionality, individual components, component integration, and manufacturability, to name a few. Current work addresses individual component development and seldom addresses the majority of other factors and for eventual impact to be made on the scales envisaged it is critical to develop this roadmap. It is also critical to identify the different disciplines and expertise which will be required to successfully accomplish the goal. This is required by not only the people involved directly with the technology, but also the funders and end users so that it is clear what the advantages of such a development would be, but also clear what hurdles need to be overcome.

1.4 Priorities for New Paper-Based Diagnostics

Although low-cost and paper-based tests such as home pregnancy and glucose tests have been introduced for home testing in developed countries, it is the critical need for low-cost tests in resource-limited settings which this chapter addresses. Table 1.2 shows the various levels of the healthcare system, encompassing both the first world and resource-limited settings, together with the typical tests available and the staff available to perform these tests at the various levels. At primary and community care levels, typically only rapid diagnostic tests are available.

It may be possible to modify some of these tests into paper-based tests that can be operated without instruments to overcome some of the problems associated with introducing POC devices into primary care level. As an example, it has been shown that the GeneXpert system can feasibly be used at the primary care level, but this is not without additional financial, operational, and logistical support [24]. Staffing requirements for the GeneXpert were underestimated; 2.5 people were required in reality compared to the estimate of 1 person. In addition, twice as many four module instruments were required than estimated to process 15 samples per day. Similar issues are experienced with other equipment-based systems when used at primary care level.

As an example, various POC instruments used for HIV testing in clinics in South Africa were evaluated. Assessment of workflow, training, and test errors was reported for various POC instruments, including Pima (CD4), HemoCue (haemoglobin), Reflotron (alanine aminotransferase, creatinine), and Accutrend (lactate). Use of the Pima CD4 showed a 4.3–6% error rate of results, and the introduction of these POC instruments increased testing times, affecting workflow and staff resource requirements. Training, connectivity, and workflow aspects were highlighted as challenges, along with the need for additional consumables, introducing various cost considerations for POC testing [15].

In addition, there are other costs associated with these systems which should not be ignored. These include equipment insurance and maintenance, stock control, administration, and communication costs. Similar test limitations were reported by Ardizzoni [25] where the main limitation was inconclusive results, which

Table 1.2 Different healthcare levels with typical equipment and tests available at each

Level	Skills available	Tests available
National	Specialists	Lab-NAT, EIA
Provincial	Specialists	Lab-NAT, POC NAT, EIA
District	Technicians	POC-NAT, EIA, RDT
Primary care	Healthcare workers	RDT
Community	Community health workers	RDT

NAT nucleic acid tests, *Lab* laboratory based, *POC* point of care, *EIA* enzyme immunoassay, *RDT* rapid diagnostic test

correlated with defective modules and cartridges and staff experience. Additionally, operational and logistical obstacles were also experienced. Substantial and continuous support was required, which is not always available in resource-limited settings. Maintenance of equipment adds further costs that are not always taken into account, particularly in resource-limited settings where harsh environments are not favourable for effective equipment operation.

This is not to state that these tests have not been very successfully rolled out and implemented, but to say that there is also a compelling case for developing tests which can completely or partially eliminate these issues.

A similar case can be made for point-of-use tests for environmental applications. Microbial water analysis requires trained personnel to carry out a number of processing steps, with a result typically only obtained after 24 h [26, 27]. Tests that can be performed rapidly on site would provide quicker feedback to users regarding contamination of water sources, ultimately saving lives. In addition, response times of water treatment works to contamination issues could also be accelerated.

Paper-based microfluidic devices provide a favourable platform on which to develop solutions for point-of-need water quality monitoring, as lightweight paper tests can be transported easily to water sources at remote locations for testing, including rivers, dams, and water plantations. In so doing, the cost of testing is substantially lowered, with testing taking place on site and without water samples having to be transported back to centralized laboratories. With lowered costs and quicker turnaround times, water quality testing can be carried out more frequently. This in turn reduces potential illness or death as water quality issues can be identified and addressed in the early stages, with far-reaching implications to maintaining and promoting the economy.

In addition to the clear impact that paper-based health and water quality diagnostics could have, particularly for resource-limited settings, a number of other useful applications could be realized. Paper tests could enable more frequent testing in both resource-limited and remote settings for food security and environmental diagnostics and enable access to contaminated sources to be prevented and controlled. Heavy metal detection and bacterial and viral tests could also be realized with paper-based tests, as well as various veterinary tests.

Developed countries have made huge strides in terms of improved healthcare and environmental consciousness. Prevalence of diseases such as diabetes and cancer is increasing at a more rapid rate in resource-limited settings, and diagnostic solutions will be critical to provide long-term solutions.

There are a great many diagnostic needs in resource-limited settings which have yet to be addressed, and this also gives opportunity for continued development, now possibly with a focus on paper diagnostics. Table 1.3 shows a number of these for presented conditions or syndromes. Table modified from Peeling et al., 2010 [28].

It may then be useful to focus on tests which can feasibly be converted from their current formats directly onto paper-based tests. From Table 1.3, the most feasible tests, and the most compelling in terms of universal demand, low cost,

Table 1.3 Unmet needs for which paper-based diagnostics may add value

Syndromes/conditions	Current diagnosis	Unmet need for POC test
Acute lower respiratory infections	Syndromic management using integrated management of childhood illness algorithms	Test/biomarker to distinguish between bacterial and viral pneumonia
Febrile illness in children	Presumptively treat for malaria in areas of high endemicity	Multiplex POC test for common causes of fever or for host biomarkers to guide treatment
Sexually transmitted infections, including HIV	Syndromic management for patients presenting with symptoms; POC tests to screen for HIV and syphilis: POC test for paediatric diagnosis of HIV; POC test for CD4 and viral load; self-tests for HIV and HCV	POC test for genital chlamydial and gonococcal infections
Antenatal care	Dual-POC test for HIV and syphilis; haemoglobin POC test for anaemia; POC test for malaria; urine test for pre-eclampsia	Multiplex POC blood-based test for screening HIV, malaria, syphilis, and anaemia and a urine-based test for pre-eclampsia

and sustainability, and therefor offering opportunity for good business models to be developed, would be:

- Diagnostics for prenatal care packages
- Simple rapid tests to distinguish between bacterial and viral infections using blood-based host biomarkers such as C-reactive protein
- Diabetes testing
- Self-tests for HIV and hepatitis C virus (HCV)

1.5 Business Models and the Economics in Developing Countries

In looking specifically at paper-based diagnostics for resource-limited settings, it is important to understand the economics of these settings. Typically in these settings, government or donors will pay for the diagnostic tests, not the patients. Most patients can't afford and thus do not have private medical insurance and rely on funds provided by the government to cover medical costs. As the numbers of these patients increase as a result of population growth and unemployment, often leading to poverty, the financial burden on governments in these countries increases.

Potential business models for paper-based diagnostics can be identified by utilizing studies which have already been done, and suggesting possible models to be considered. It would appear that global aid agencies would still be required to play a central role in providing funding.

Without a critical look at how these technology developments and implementations will be funded, impact in resource-limited settings will be limited. A number of studies have been done to determine the impact of various interventions within the healthcare community. Of particular interest to this work is the impact which paper-based diagnostic devices could have on the healthcare funding models and in resource-limited environments.

A recent study of community health workers (CHWs) and their impact on the health status of the population showed that CHWs do have the potential to improve healthcare, particularly for those disadvantaged or living in hard-to-reach areas [29]. Although it is difficult to predict cost savings of paper-based diagnostics, it is precisely these areas where they would show most benefit, as these areas typically do not allow for more advanced testing with equipment-based systems. Specific interventions to be focused on would include diabetes, HIV/TB, hypertension, and mother and child health, and here there are already a number of groups working on developing paper-based diagnostic tests.

A study of the GeneXpert system in 2015 concluded that its implementation in clinical settings led to an increase in confirmed TB cases, for use both as a first test and as an add-on, and that it could be feasibly implemented [25]. However, it also concluded that costs should be carefully considered, that the system was not a plug-and-play device, and that significant additional requirements were necessary (infrastructure, training, technical support, and experience). As a final statement, the authors state that in order to further decentralise diagnosis of TB, more robust, simpler technologies which are adapted to resource-limited settings are required. This would further suggest a role to be played by paper-based diagnostics.

Similar studies looked at estimating the cost of point-of-care CD4 testing utilising the Alere Pima analyser in South Africa [30]. Extensive analysis concluded that implementing these tests at the point of care rather than at a more central laboratory substantially increased the cost per test (up to three times more expensive). The business case for point-of-care diagnostics for resource-limited settings has also been summarized with focus on HIV testing, highlighting the challenges in getting products successfully into the market [31].

These studies all utilise existing diagnostics, and it is in many ways intuitive that costs per test would be substantially more expensive in a clinical environment, as cost items such as insurance, training, low throughput, maintenance, and transport are all expected to be more expensive than at centralised laboratories. However, development of more advanced paper-based diagnostics is expected to eliminate many of these hurdles and costs, so that items such as insurance and transport will essentially be eliminated, and items such as training will be reduced substantially. From this perspective it would perhaps be insightful to develop a fuller analysis of costing which would be achieved with an ideal ASSURED device.

Models developed should also not only follow a cost-based criteria, as factors such as usability, time to result, accuracy, maintenance, and specialised skills required should all be taken into account. Improvement in these aspects may provide great value to the overall impact and business case for paper-based diagnostics.

However, it is envisaged that the actual cost of the device or test will be substantially less than equivalent equipment-based systems, and it should be possible to reduce other costs related to use of tests at the point of care. The main cost-saving aspects introduced by paper-based point-of-care diagnostics are listed below.

- Paper-based test devices themselves will be cheap to manufacture, as conventional and well-understood manufacturing methods will be employed.
- Multiplexing of these tests is possible.
- Material costs will be low.
- The technology is massively scalable, which will further allow for reduced costs.
- Level of operational skill required, and thus training, is reduced since the end goal is for these tests to be more easily usable than conventional tests.
- Minimally instrumented or equipment-free solutions, reducing dependency on electricity and other infrastructural elements, and also eliminating maintenance requirements.
- Colorimetric tests (and possibly other tests where mobile phone readout is possible) allow for the possibility of remotely analysing results via telemedicine. In this way, expertise is available when required.

Diagnostics for All (DfA, www.dfa.org), a spin out from the George Whitesides laboratory at Harvard University, has as its mission "*to save lives and improve health in the developing world through pioneering technological innovation*". They have focused on paper-based diagnostic systems, and also suggested a business model. The model is based on selling diagnostics to the first world, and then using the proceeds to help the third world. Conventionally, companies sell high-margin products to generate profits, and then donate a small percentage of the profit to needy sources. DfA would use the same approach for their first-world customers, but then plough all profit back into providing diagnostics and training to the third world. Initial focus areas would include liver function tests, child nutrition, nucleic acid-based testing, and farmer support.

Some examples of paper-based diagnostic devices that are in the market from Diagnostics for All include three-dimensional μPADs for liver testing, nutrition testing, and nucleic acid detection. Other paper-based devices in the market include the text-reported blood typing test from Haemokinesis and a lateral flow immunoassay test for infectious disease testing by INSiGHT. Various challenges remain in the ongoing development and improvement of these devices, including reagent stability, need for external equipment (incubators, readout), and data connectivity, which have been summarized [6]. The business models for these devices typically consist of collaborations with both universities for the technical development and commercial for-profit partners.

In order for these types of business models to work, large-scale partnering will be required. This will be necessary with aid organisations that already have extensive networks, but also with conventional diagnostics suppliers and drug manufacturers.

As an example, it may be possible to package home- or clinic-based diagnostic tests with the drug of interest so that continual monitoring can take place.

Paper-based diagnostic devices may not provide the entire solution, but may form part of a greater solution to alleviate the burden of various costs associated with current point-of-care diagnostic solutions. In this regard, successful implementation and development of paper-based diagnostic devices would feed into the economic models which have been developed for these areas. There would be a positive benefit in terms of increased employment, increased productivity, and also multiplier effect of new money put into the economy. This is estimated to be approximately 1.5× [29]. As part of a greater solution, paper-based diagnostic devices utilized as screening tools have great potential. Where current limitations such as sensitivity and specificity as well as complete automation of devices hinder the implementation of truly ASSURED paper-based devices, invaluable tests for screening or to be used in conjunction with other equipment-based tests can be developed on paper. The use of these types of tests should not be understated, as this type of test structure is commonly and effectively utilized in resource-limited clinics. For example, patient questionnaires are used for TB screening, and rapid HIV tests are used to screen patients before a blood sample is sent for laboratory analysis in numerous South African clinics. This emphasises that although there may not currently be a perfect paper-based point-of-care diagnostic device, these tests may still provide useful results, results that are "good enough", be it for screening or rule-in and rule-out scenarios [32].

In order to determine the value of these diagnostics, an attempt has also been made to introduce specific frameworks to determine value. As an example, value can be determined as the performance × efficiency of the diagnostic. Here, performance is given by the technical accuracy of the test over the turnaround time, and efficiency is given by utility (% of decisions made) over costs[3]. Similar value determinations could be made for resource-limited environments in determining the value of introduced diagnostics.

The WHO CHOICE program (http://www.who.int/choice/about/en/) is aimed at identifying the healthcare interventions which provide the most value for money. Some of these interventions may only require screening tests or tests that do not have to be fully quantitative or extremely accurate, and paper-based diagnostics could thus address both value for money and lower device costs. In future, as paper and printing technologies advance, so too will paper-based diagnostics, enabling direct diagnosis to be performed using these devices.

1.6 Paper-Based Diagnostics: The Big Picture

This chapter has discussed the potential of paper to fulfil the ASSURED criteria for developing point-of-care diagnostics for resource-limited settings.

Clearly there is a need for point-of-care diagnostics to be developed to fill these gaps, and paper-based components and functionalities could play a role in achieving

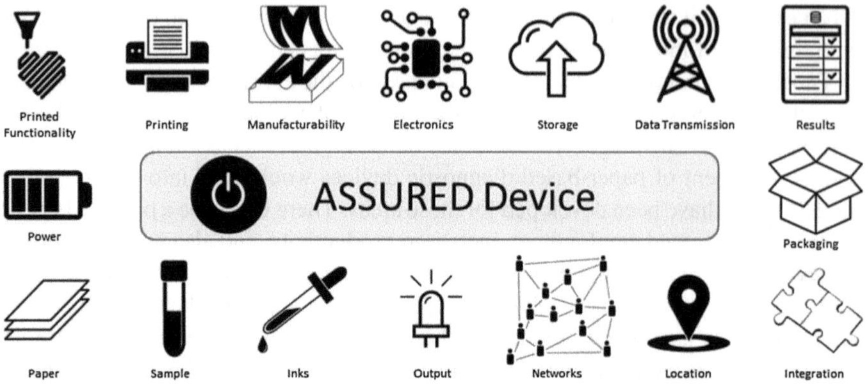

Fig. 1.6 Various components which need to be further developed and integrated to form the ideal test. Graphics: The Noun Project: Bohdan Burmich, Arafat Uddin, Sarah Marquez, Jhun Capaya, Michael Wohlwend, DewDrops, AA, Dant, Icon Fair, Vectors Market, Alvarobueno, Alex Muraviov, Creative Stall, Santiago Arias, Marie van den Broek, Hunotika, Gregor Cresnar, Bonegolem, Michele Zamparo, Andrew Hainen, Ralf Schmitzer, Arthur Schlain, Alex Muravev, Rockicon

this. This would allow for sophisticated lab equipment processes to be brought down the pyramid (Fig. 1.2) from urban to rural/clinic settings. Equipment-based diagnostics typically implement processes such as PCR, molecular diagnostics, and detection methods, typically based on colorimetric, fluorescence, and absorption techniques. These processes are essential in many diagnostic tests and are also applicable to addressing the unmet needs presented in Table 1.3.

A number of these processes have been implemented to some extent on paper devices, including DNA extraction with on-chip reagent storage [33], DNA detection [34], and loop-mediated isothermal amplification and detection integrated on paper devices [35, 36]. Molecular diagnostic platforms on paper have also been implemented [37, 38]. For many of these platforms, a degree of skill is required as a number of steps are manual and require user interaction. In many cases external instrumentation is also required, requiring power and increasing costs and operational complexity of the platform.

For components that are still missing or that require further automation or optimization, various paper-based and printed components may contribute towards realizing fully ASSURED paper-based point-of-care diagnostics. Figure 1.6 summarizes the various functional components that have been and continue to be developed on paper.

As an example, for fully automated and integrated PCR to be implemented on paper, more advanced heating set-ups would be required, which would stem from the development of printed functional inks (e.g. PTC inks), as well as printed electronic components for accurate temperature control. Heating on paper [39] as well as for application in paper-based diagnostics has been explored [40].

Quantitative outputs can also be implemented on paper to enhance the usual yes/no answer obtained from lateral flow devices. Examples of current paper-based diagnostic tests that utilize fluorescence or colorimetric readout using portable handheld instrumentation or mobile phones have also been implemented [6]. Towards realizing fully ASSURED paper-based devices, integrated sensors on paper [41], including electrochemical, impedance, pressure, gas, and light sensors, can be implemented, as well as data processing [42] and displays [6] on board the device. Integration of printed electronic components in terms of colour detection mechanisms and processing components, as well as displays or outputs to provide feedback, would be required for effective, integrated paper-based diagnostics to be achieved.

A number of printed display technologies have already been implemented on paper, from early thermochromic displays [43] to more recent light-emitting paper examples [44]. The result can be stored on the device and transmitted through wireless communication techniques, where printed antennas on paper can readily be employed. Additionally, automation of user steps could be achieved using paper-based actuators in the form of valves and switches, which can be activated by the introduction of the sample to the device.

Future development of components for paper-based diagnostics could benefit from a scoring system for developers of these devices, where the peripheral functionalities or even the entire test could be scored. Rating the various functional blocks according to the ASSURED criteria and other important considerations required for resource-limited settings and relevant clinical outcomes could provide a solid foundation on which to build future successful paper-based diagnostic devices. These scores could then be combined with the diagnostic accuracy or performance of the test to calculate the "test efficacy" [45].

With the advances in printed functionality and various integrated examples highlighting the progress made in the field of point-of-care paper-based diagnostics, it remains to be seen whether these solutions will indeed improve patient outcomes. As solutions are developed and deployed in real-world settings, extensive evaluations will need to be carried out to assess whether the patient receives improved or more rapid treatment as a result of on-site testing. Certainly, interest has grown significantly for point-of-care diagnostics, and seems to set to continue, and we have tried to identify the critical need for paper-based diagnostic tests while also setting goals which developers can aim for going forward.

References

1. The Lewin Group et al (2005) The value of diagnostics innovation, adoption and diffusion into health care. Health Care (Don Mills) 214. https://dx.advamed.org/sites/dx.advamed.org/files/resource/Lewin%20Value%20of%20Diagnostics%20Report.pdf. Accessed 22 Sept 2018
2. The Lewin Group (2009) The value of laboratory screening and diagnostic tests for prevention and health care improvement. https://thehill.com/sites/default/files/acla_lewinvalueoflabsxdxreport_0.pdf. Accessed 23 Sept 2018

3. Rohr U-P et al (2016) The value of in vitro diagnostic testing in medical practice: a status report. PLoS One 11:e0149856
4. Nowak R (1995) Rockefellers big prize for STD test. Science 269:782
5. Mabey D, Peeling RW, Ustianowski A, Perkins MD (2004) Diagnostics for the developing world. Nat Rev Microbiol 2:231–240
6. Yamada K, Shibata H, Suzuki K, Citterio D (2017) Toward practical application of paper-based microfluidics for medical diagnostics: state-of-the-art and challenges. Lab Chip 17:1206–1249
7. Urdea M et al (2006) Requirements for high impact diagnostics in the developing world. Nature 444(Suppl):73–79
8. Ferrer A, Pal L, Hubbe M (2017) Nanocellulose in packaging: advances in barrier layer technologies. Ind Crop Prod 95:574–582
9. Hoeng F, Denneulin A, Bras J (2016) Use of nanocellulose in printed electronics: a review. Nanoscale 8:13131–13154
10. Khwaldia K, Arab-Tehrany E, Desobry S (2010) Biopolymer coatings on paper packaging materials. Compr Rev Food Sci Food Saf 9:82–91
11. Cheng B et al (2016) Data connectivity: a critical tool for external quality assessment. Afr J Lab Med 5:535
12. Gous N et al (2018) Expert review of molecular diagnostics the impact of digital technologies on point-of-care diagnostics in resource-limited settings. Expert Rev Mol Diagn 18:385–397
13. Thom NK, Yeung K, Pillion MB, Phillips ST (2012) "Fluidic batteries" as low-cost sources of power in paper-based microfluidic devices. Lab Chip 12:1768
14. Nguyen TH, Fraiwan A, Choi S (2014) Paper-based batteries: a review. Biosens Bioelectron 54:640–649
15. Stevens W, Gous N, Ford N, Scott LE (2014) Feasibility of HIV point-of-care tests for resource-limited settings: challenges and solutions. BMC Med 12:173
16. Kumar AA et al (2015) From the bench to the field in low-cost diagnostics: two case studies. Angew Chem Int Ed 54:5836–5853
17. Piila T, Juusti A, Kylmanen M, Ikonen K (2015) Printed electronics & diagnostic products printocent designer's handbook. Neficon Finland Oy, Haukipudas
18. Anstey Watkins JOT, Goudge J, Gómez-Olivé FX, Griffiths F (2018) Mobile phone use among patients and health workers to enhance primary healthcare: a qualitative study in rural South Africa. Soc Sci Med 198:139–147
19. Oldach R (2015) Pioneering new diagnostics: addressing challenges and implications for point-of-care testing in African settings. Lab Cult ASLM Mag 13:7–17
20. Clarke SF, Foster JR (2012) A history of blood glucose meters and their role in self-monitoring of diabetes mellitus. Br J Biomed Sci 69:83–93
21. Farrell BO (2013) Lateral flow immunoassays systems: evolution from the current state of the art to the next generation of highly sensitive, quantitative rapid assays. The immunoassay handbook. Elsevier Ltd.
22. O'Farrell B (2015) Lateral flow technology for field-based applications-basics and advanced developments. Top Companion Anim Med 30:139–147
23. Yetisen AK, Akram MS, Lowe CR (2013) Paper-based microfluidic point-of-care diagnostic devices. Lab Chip 13:2210
24. Clouse K et al (2012) Implementation of Xpert MTB/RIF for routine point-of-care diagnosis of tuberculosis at the primary care level. South African Med J 102:805–807
25. Ardizzoni E et al (2015) Implementing the Xpert1MTB/RIF diagnostic test for tuberculosis and rifampicin resistance: outcomes and lessons learned in 18 countries. PLoS One 10:1–15
26. Hamilton WP, Kim M, Thackston EL (2005) Comparison of commercially available Escherichia coli enumeration tests: implications for attaining water quality standards. Water Res 39:4869–4878
27. Rompré A, Servais P, Baudart J, De-Roubin MR, Laurent P (2002) Detection and enumeration of coliforms in drinking water: current methods and emerging approaches. J Microbiol Methods 49:31–54

28. Peeling RW, Mabey D (2010) Point-of-care tests for diagnosing infections in the developing world. Clin Microbiol Infect 16:1062–1069
29. Daviaud E, Budlender D (2017) Saving lives, saving costs: investment case for community health workers in South Africa. What costs and what benefits. For the health sector. For the economy and For society? South African Medical Research Council, Cape Town, pp 1–81. http://www.mrc.ac.za/sites/default/files/files/2017-10-30/SavingLivesSavingCosts.pdf. Accessed 23 Sept 2018
30. Larson B et al (2012) How to estimate the cost of point-of-care CD4 testing in program settings: an example using the Alere Pima™ analyzer in South Africa. PLoS One 7:e35444
31. Drain PK, Rousseau C (2017) Point-of-care diagnostics: extending the laboratory network to reach the last mile. Curr Opin HIV AIDS 12:175–181
32. Weyer K, Carai S, Nunn P (2011) Viewpoint TB diagnostics: what does the world really need? J Infect Dis 204:S1196–S1202
33. Tang R et al (2017) A fully disposable and integrated paper-based device for nucleic acid extraction, amplification and detection. Lab Chip 17:1270–1279
34. Qiu X et al (2017) Instrument-free point-of-care molecular diagnosis of H1N1 based on microfluidic convective PCR. Sensors Actuators B Chem 243:738–744
35. Rodriguez NM et al (2015) Paper-based RNA extraction, in situ isothermal amplification, and lateral flow detection for low-cost, rapid diagnosis of influenza a (H1N1) from clinical specimens. Anal Chem 87:7872–7879
36. Seok Y et al (2017) A paper-based device for performing loop-mediated isothermal amplification with real-time simultaneous detection of multiple DNA targets. Theranostics 7:2220–2230
37. Choi JR et al (2015) Paper-based sample-to-answer molecular diagnostic platform for point-of-care diagnostics. Biosens Bioelectron 74:427–439
38. Pardee K et al (2014) Paper-based synthetic gene networks. Cell 159:940–954
39. Fischer T, Wetzold N, Elsner H, Kroll L, Hübler AC (2011) Carbon nanotube areas—printed on textile and paper substrates. Nanomater Nanotechnol 1:3
40. Phillips EA, Shen R, Zhao S, Linnes JC (2016) Thermally actuated wax valves for paper-fluidic diagnostics. Lab Chip 16:4230–4236
41. Mattana G, Briand D (2016) Recent advances in printed sensors on foil. Mater Today 19:88–99
42. Chang J, Zhang X, Ge T, Zhou J (2014) Fully printed electronics on flexible substrates: high gain amplifiers and DAC. Org Electron Physics Mater Appl 15:701–710
43. Siegel AC, Phillips ST, Wiley BJ, Whitesides GM (2009) Thin, lightweight, foldable thermochromic displays on paper. Lab Chip 9:2775
44. Asadpoordarvish A et al (2015) Light-emitting paper. Adv Funct Mater 25:3238–3245
45. Drain PK et al (2014) Diagnostic point-of-care tests in resource-limited settings. Lancet Infect Dis 14:239–249

Chapter 2
The Role of Paper Chemistry and Paper Manufacture in the Design of Paper-Based Diagnostics

Samuel Schabel and Markus Biesalski

Abstract Paper has been used intensively for thousands of years, and it is a material made from renewable resources. It consists of a highly porous, bendable and foldable, flat structure of randomly arranged and connected fiber-like basic building blocks. The possibility to transport fluids without pumps and sophisticated dosing systems is attractive, and microfluidic paper for diagnostic applications has gained increasing interest, in particular, throughout the last decade. Although a number of interesting demonstrator devices for such easy-to-use diagnostic systems have been reported, still only a very limited number of devices made it into the market. The latter is mainly caused by the geometric and chemical complexity of the paper material itself. Whereas chemical functionalization (e.g., for defining hydrophobic barriers for spatially resolved fluid transport) is well advanced and will be covered by other groups in this book, understanding the impact of the paper material itself on the performance of paper-based diagnostic devices is still a challenge. Yet, only if we understand the latter from a fundamental point of view, more advanced and successful paper materials for such applications may become available. This book chapter outlines principles of paper manufacture; it thereafter reviews important paper chemistry aspects, and finally highlights recent developments on how engineering the morphology and chemistry of paper sheets gives us more insight into fundamentals on the role of the material itself in paper-based diagnostic applications.

S. Schabel
Institute for Paper Technology and Mechanical Process Engineering, Technische Universität Darmstadt, Darmstadt, Germany
e-mail: schabel@pmv.tu-darmstadt.de

M. Biesalski (✉)
Institute for Macromolecular Chemistry and Paper Chemistry, Technische Universität Darmstadt, Darmstadt, Germany
e-mail: biesalski@tu-darmstadt.de

K. J. Land (ed.), *Paper-based Diagnostics*,
https://doi.org/10.1007/978-3-319-96870-4_2

2.1 Paper Manufacture

Paper is a material based on natural resources. According to standards like DIN 6735, paper is specified as a flat material consisting mainly of natural fibers, which are formed to a flat material by spreading a fiber suspension on a wire. Many different types of natural fibers are used for papermaking. In conventional papermaking softwood and hardwood trees like spruce, pine, eucalyptus, and birch are dominating [1]. In countries with low forest areas papermaking from annual plants like wheat or rice straw is common.

For special paper applications like filters for industrial, medical, or home applications or for banknote paper, cotton and cotton linters are commonly used. Although this raw material is more expansive as compared to wood fibers, cotton and cotton linters have some advantages with respect to chemical composition: here almost no lignin and hemicelluloses are present and the fibers do consist of almost exclusively type I cellulose.

Due to the natural origin of the papermaking raw material, papermakers have to handle high variation of properties like fiber length, fiber width, color, strength, and flexibility. However, it is exactly this wide variation of paper constitution, which allows for a large area of typical paper applications. At present (i.e., in 2016 and 2017), more than 400 million tons of paper are being produced worldwide per year [2]. Major applications of paper are packaging materials (about 50%), graphical paper (about 40%), and so-called specialty paper (about 6%) [2]. Whereas graphical paper-grade production is decreasing due to the digital revolution, production of paper materials for packaging and specialties is increasing. With respect to the latter, a recent market study expects a very strong market growth with such specialty papers from 11 billion US$ in 2014 to more than 26 billion US$ in 2022 [3]. This increase is partly due to novel and advanced paper-based applications in various areas, including medical diagnostics. However, microfluidic paper applications as of today are still a small market for paper producers. Except experimental studies and setup in research organizations, only paper-based test strips for pH determination or determination of the blood glucose level in urine can be found on the market, and the amount of paper needed for this type of applications is quite small, compared to the total world production of paper.

For most of such microfluidic papers, up to now, cotton cellulose-based papers like filter papers were used due to the relatively small variations of properties compared to other paper raw materials without extra treatment. The latter is mainly due to the chemical and morphological composition and structure of the cotton linter fibers. However, as becomes evident when overseeing the present scientific literature with microfluidic papers, researchers do not consider at all differences in the chemical and geometrical paper structure nor do they consider that during paper manufacture a number of additives are being used and some of them may still be part of the final filter paper. As such, our knowledge of how paper intrinsic properties affect the performance of microfluidic paper is still very poor. This chapter is therefore dedicated to first provide the reader of this book on paper-based diagnostics with some general knowledge on pulping and paper manufacture (Sect. 2.1), and in addition the chemistry of papermaking and chemical additives being used in paper production (Sect. 2.2). Both areas will be briefly discussed with the focus on relevant processes for microfluidic papers. Note that a comprehensive overview on pulp and paper production can be found in various textbooks [4] and will not be covered in complete here.

Papermaking can be divided into two subprocesses, which are briefly discussed within the following sections:

- The manufacturing of papermaking fibers from wood or other raw materials (pulping)
- The manufacturing of the fiber web (papermaking)

2.1.1 Pulping and Processing of Fibers

For producing fibers from natural material like wood or straw two major processes are used in industry. In chemical pulping wood or wood chips are cooked in solvents which dissolute the lignin as substance which is gluing together the natural fibers in wood. By this dissolving process single fibers are relieved. In chemical pulping around 40% of the raw material of wood is dissolved and the remaining about 60% consisting of fibers can further be processed. Typically, in today's pulp mills the dissolved organic material is used as biofuel to fire the pulping process and produce exhaust energy for electricity supply, mostly. The properties of the fibers produced in this way depend strongly on the chemicals used for pulping, the process conditions and design, and the posttreatment like bleaching. Detailed information about chemical pulping can be found in a textbook [5].

A second method to produce fibers from wood is mechanical pulping where logs or wood chips debarked are treated mechanically in grinding or refining machines and single fibers are broken out of the wood matrix with mechanical forces. This mechanical process typically has a yield of around 90% and produces fibers still mantled with lignin which gives those fibers mechanically and chemically different properties compared to chemical pulp fibers. More information about mechanical pulping can be found in a textbook [6]. After producing the fibers for typical paper-making applications additional treatments can follow for cleaning the pulp from impurities and for mechanical modification of the fibers. For separating impurities screens, hydro-cyclones and partly flotation are used. Detailed descriptions about these processes and treatments can be found here [7]. The most important process for microfluidic applications of paper is the so-called refining or beating process. For refining fibers are processed in machines with two refining plates, each with a notched pattern on its surface passing a small gap so that high shear of the fast-rotating tools is applied to the surface of the fibers. This process is operated with fibers in the form of a suspension with a consistency between 3 and 20%, typically.

Figure 2.1 shows an image of a typical refiner tool. Depending on the notched pattern, the height of the channels, the gap width, and the speed of the refiner plates

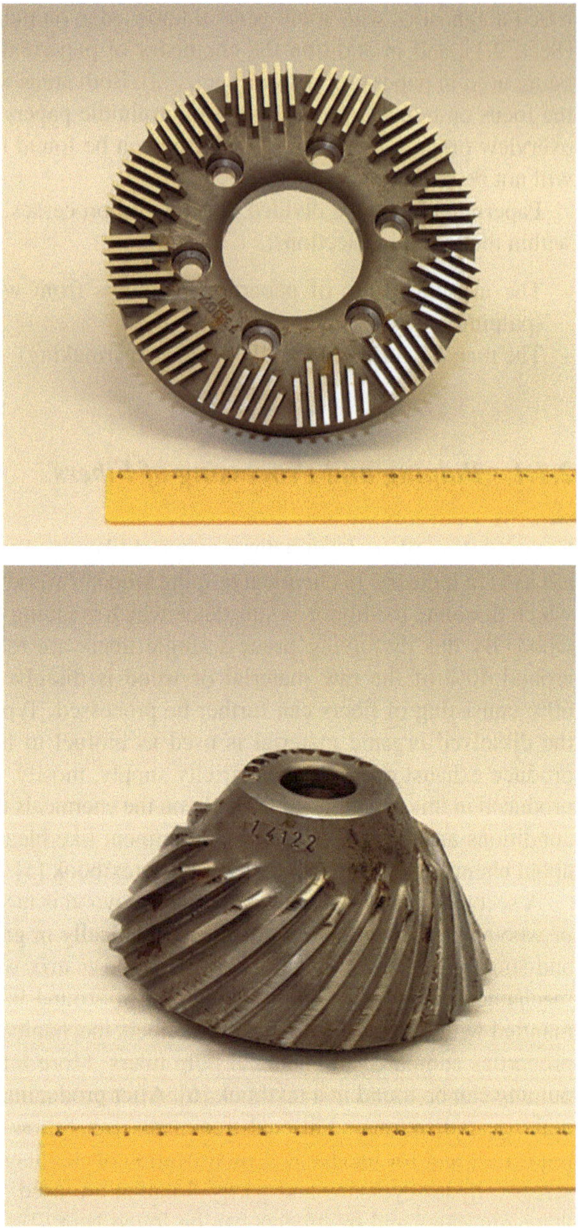

Fig. 2.1 Images of typical refiner plates (disc refiner—top; conical refiner—bottom)

(typically circumferential velocities are about 20 m/s), fibers can be fibrillated, curled, or even cut into pieces. By this the refining step is very efficient in producing fiber fractions with for instance defined morphological properties like length. The refining step gives a very strong effect on the porosity of the paper produced. Such

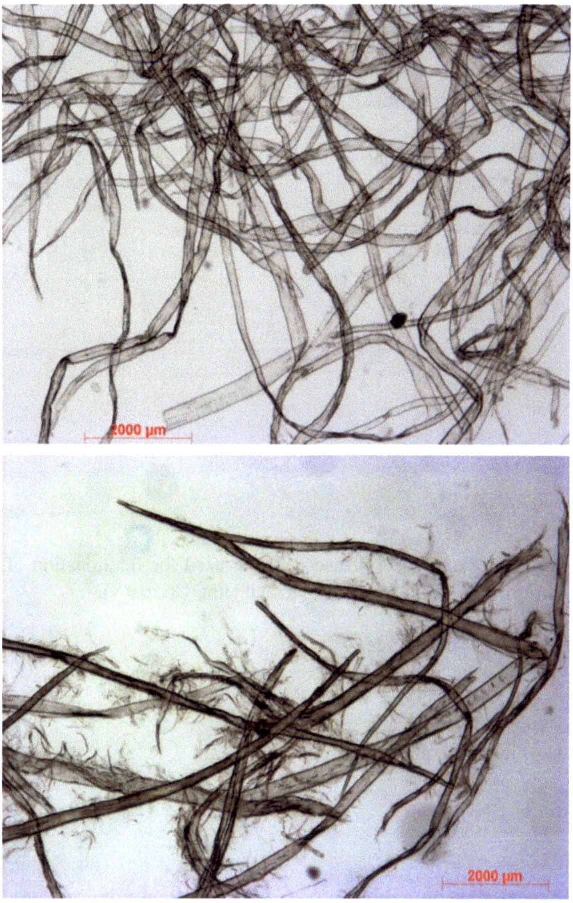

Fig. 2.2 Pine kraft pulp fibers before refining (top) and after 90-min treatment in a Jokro mill (bottom)

refining devices are available in lab, in pilot, and in industrial scale. Figure 2.2 shows microscope images from fibers before and after refining treatment.

Refining of course has a high impact on the porosity of paper and therefore also on the transport of fluids in a paper by capillary forces. A widely used method to characterize the degree of refining is the drainability according to Schopper-Riegler (ISO 5267-1) and the transport of fluids by capillary forces can be characterized by the capillary rise value according to Klemm-Test ISO 8787. Besides refining, fractionation is a very common unit operation for designing the fiber morphology. While refining changes the fiber length, and surface structure, fractionation is a process for classifying a mixture of fibers in specific size classes. Since paper is typically made from natural materials the fiber dimensions are not all similar and processes have to operate with a distribution of fiber dimensions like length, thickness, and wall thickness. The most widely used machines for fractionation of pulp fibers are so-called *pressure screens*. Pressure screens consist of a cylindrical or conical housing, a cylindrical screen basket, and a rotor as shown in the sketch in Fig. 2.3.

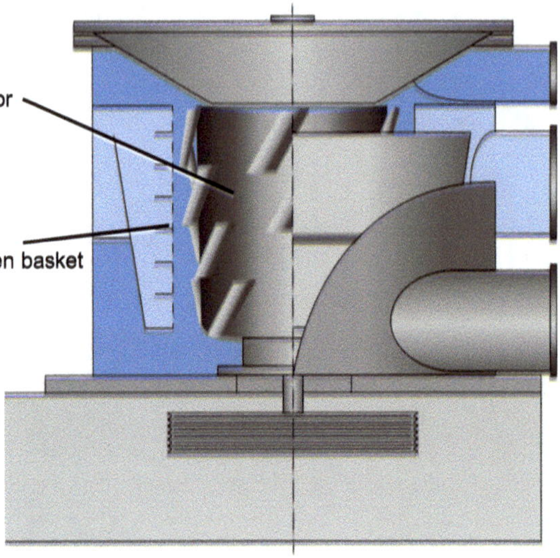

Fig. 2.3 Sketch of a pressure screen used for fractionation of fiber suspensions (source: Voith GmbH & Co. KGaA, 89522, Heidenheim, Germany)

Fig. 2.4 Typical screen baskets (source: Voith GmbH & Co. KGaA, 89522, Heidenheim, Germany, HeculeX C-bar basket)

The most important element defining the fractionation effect is the screen basket. The basic design of such a screen basket is shown in Fig. 2.4. The screen basket consists of slots of a defined width which is typically in the range of 100–250 μm for fractionation of pulp fibers. In order to force pulp fibers to pass through slots of such a small size the surface of the screen basket has to be formed in such a way

Fig. 2.5 Visualization of flow around pressure screen openings (source: Voith GmbH & Co. KGaA, 89522, Heidenheim, Germany)

that microturbulence is created and the vortices pull the fibers through the slots. Depending on the width of the slots the fibers can be fractionated mainly by length but also other properties like flexibility and specific surface of the fibers play a role in fractionation.

Alternatively, screen baskets with holes in the diameter range of 0.8–2 mm can be used for fractionation. Figure 2.5 illustrates the typical flow through slots of a slotted screen basket. For this picture the flow is visualized by tracer particles and a laser light sheet method. The vortices formed by the triangular outline contour of the slot can be seen in the picture.

2.1.2 *Papermaking*

Papermaking for microfluidic applications can be done in industrial scale but also in lab scale, since for this kind of applications often very small amounts and small pieces of paper sheets are used. The production output of a typical industrial paper mill is in the range of 500 tons per day to 1000 tons per day. It can be as high as 4000 tons per day [8]. For microfluidic application usually much smaller amounts are required and therefore such papers are often produced in lab or pilot-scale machines. In the following sections we outline both papermaking in the lab scale and pilot scale and industrial papermaking.

2.1.3 *Rapid-Köthen Handsheet Method*

For lab sheet formation, the Rapid-Köthen method is most common in Europe. The process is described in EN ISO 5269-2, and Fig. 2.6 displays a typical Rapid-Köthen instrument used in the lab. With this method, typically circular sheets are

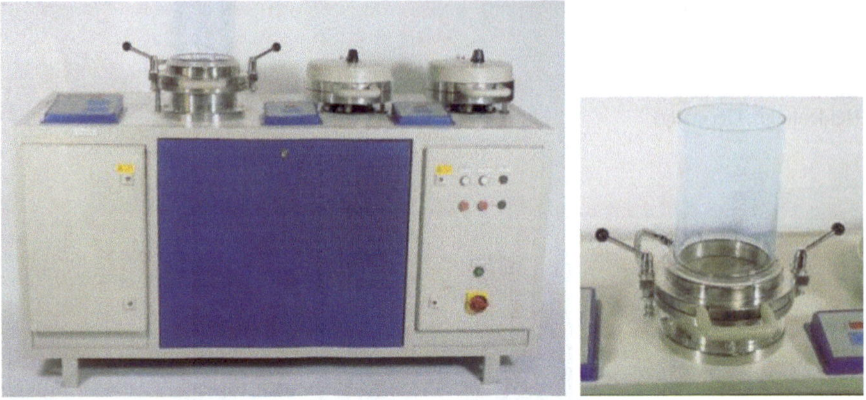

Fig. 2.6 Rapid-Köthen device for laboratory handsheet manufacturing

formed with a diameter of about 20 cm. The minimum grammage is about 40 g/m^2, depending on the fiber type. The sheets typically have a high degree of homogeneity with respect to fiber distribution and local variations of the grammage (i.e., sheet density). The fibers are oriented randomly in the plane of the sheet. The drying of the sheets is under conditions simulating typical paper machine drying (refer to cylinder drying in Sect. 2.1.5) with a moderate pressure and vacuum.

Due to the pressure and vacuum application during drying, shrinkage of fibers is restrained, which is common for industrial paper drying. Since papermaking fibers have a high swelling and shrinking potential, shrinkage-restrained drying results in significant "frozen tensions" in the sheets. The fibers can double their diameter by swelling but their length is almost unchanged. During drying the diameter reduces to the initial size but now the fibers are included in a fiber mat and hydrogen bonds have formed. Thus, the drying induces mechanical tension and stresses which affect the mechanical properties of the sheet [9].

2.1.4 Dynamic Handsheet Forming

Since the Rapid-Köthen method produces handsheets with a uniform fiber orientation the so-called dynamic handsheet-forming process can be used for the production of lab handsheets with a fiber orientation comparable to industrial paper machines. Figure 2.7 schematically illustrates the setup of a dynamic handsheet-forming process. The main element of this device is a rotating drum like in a centrifuge where a paper machine wire is spanned on a cylinder drum. Fiber suspension is sprayed on the inner surface of this wire and due to these centrifugal forces dewatering can take place. The speed difference between the circumferential speed of the drum and the jet produced by the spraying nozzles determines the orientation of the fibers. The higher the speed difference the higher the degree of orientation in sheets produced by this method.

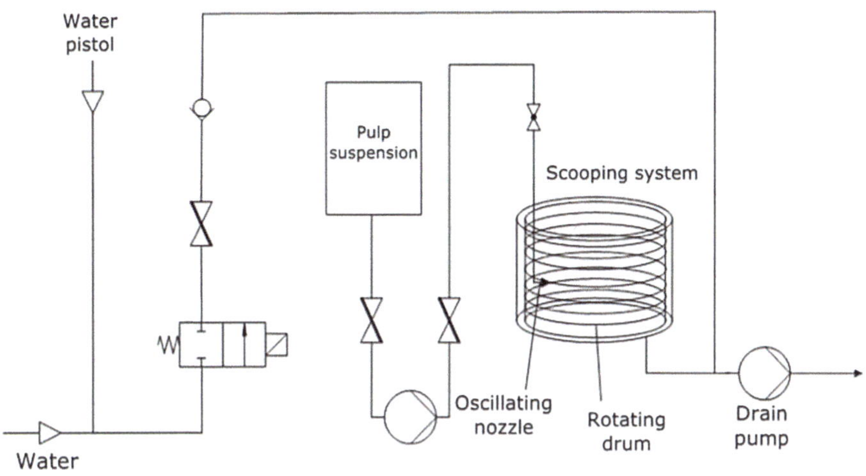

Fig. 2.7 Sketch of a dynamic handsheet-forming device

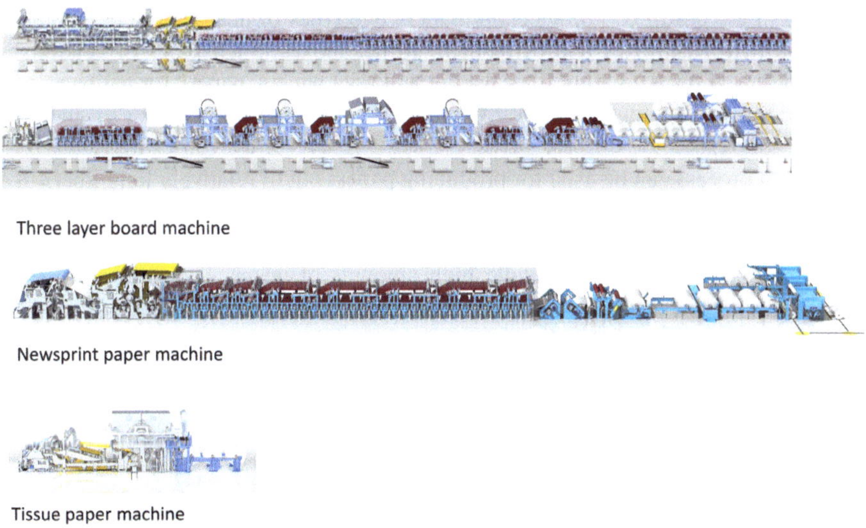

Fig. 2.8 Basic layouts of paper machines for the most important standard paper types (Source: Voith GmbH & Co. KGaA, 89522, Heidenheim, Germany)

2.1.5 Industrial Papermaking

In an industrial paper machine pulp suspension with a consistency of typically 1% is formed into a sheet and then dewatered by mechanical forces and by drying. Typical size ranges of paper machines are up to a width of 12 m and a speed of 2000–3000 m/min and productions up to more than 4000 tons per day. Figure 2.8

shows the basic components of industrial paper machines for the most important paper types, where the main components of such a paper machine consist of the following:

- Headbox and former: In this section the continuous flow of pulp suspension is distributed equally over the width of the machine and a flat rectangular jet is formed and injected on a moving wire. By this an endless and uniform sheet is formed. The thickness and uniformity of the paper are defined in this first section. Here also dewatering starts by mechanical forces and vacuum.
- Press section: After the wire has formed and has reached a consistency of about 20% dry content further dewatering is achieved by pressing between rolls. With mechanical forces including pressing a dry content of the sheet of up to 55% can be reached.
- Drying section: After pressing only thermal drying can be used to dry the paper up to an equilibrium moisture content of about 5–8%.

The most important part for the properties of a paper relevant for microfluidic applications like porosity, uniformity of mass distribution, and fiber orientation are defined in the headbox. For specialty papers two types of headboxes are common: inclined wire headboxes and cylinder mold headboxes. Such types of headboxes today are used for producing filter paper, banknote papers, papers for tea bags, and reinforcement fiber mats. Figure 2.9 shows the designed drawing for an inclined wire headbox.

Such a process can be used for producing papers out of relatively long fibers like cotton, abaca, flax, hemp, or synthetic fibers or of course mixtures of such fibers. Since the suspension has to be more diluted if longer fibers are used such paper machine headboxes have to handle high amounts of water. The dilution in such headboxes is typically below 0.5% which means that for producing 1 kg of paper at

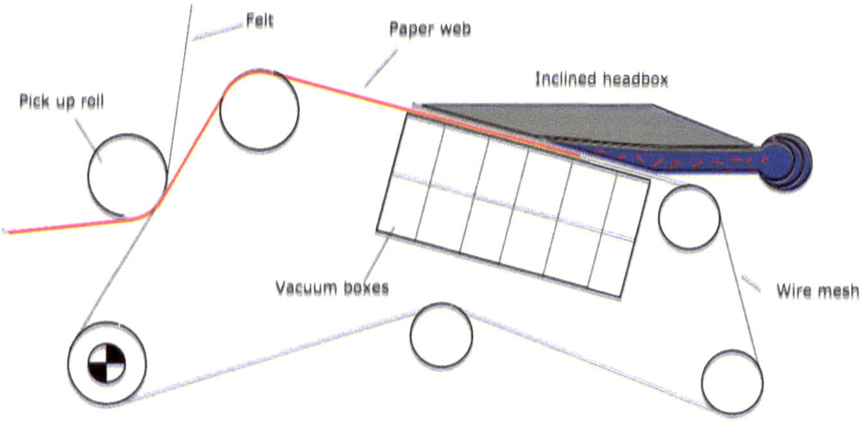

Fig. 2.9 Principle of an inclined wire headbox

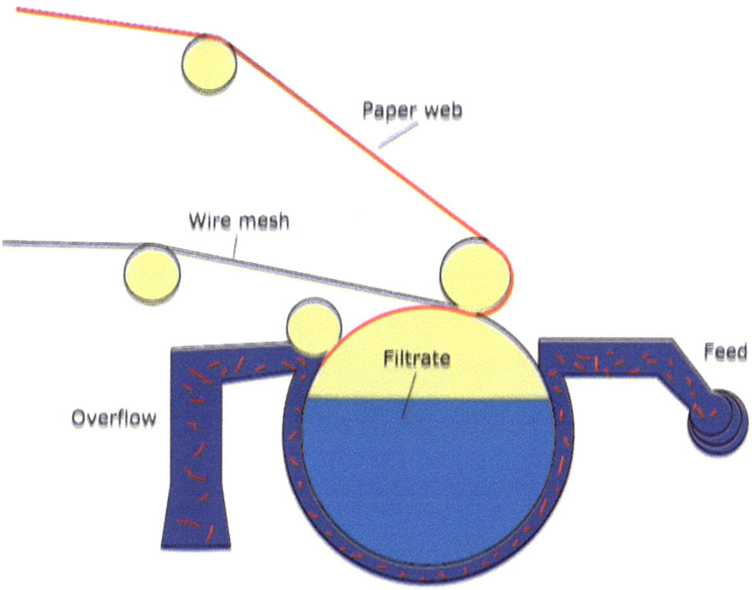

Fig. 2.10 Cylinder mold headbox

least 200 kg water has to be processed. Alternatively cylinder mold headboxes can be used. Such a headbox is schematically illustrated in Fig. 2.10. Cylindrical wire headboxes can produce papers with a relatively high thickness and with a complex three-dimensional structure as it is utilized for example banknotes with water marks or other safety elements.

After the wire section a dryness of the paper web of about 20% is reached and this is typically not high enough to have a self-supported paper web. This means that in the following section, the press section, the paper web has to be supported to avoid brakes. From the forming wire the web is transferred to a felt by a so-called pickup roll and transferred into the press section as it can be seen in the sketch in Fig. 2.11.

A typical press section consists of 3–5 nips between rolls which are pressed together. The paper web is supported by two felts (upper and lower felt) which have the function to support the paper web but also to take the water which is pressed out of the paper web. This sandwich (felt-paper web-felt) is transferred to the press nips where press rolls are pressing against each other and by this a pressing force is applied to the paper web. Very common is the so-called center roll concept, where two or more press nips are arranged around one center roll. The dewatering in this press section leads to a dry content of 40–50% (or even more in high-performance press section for some paper products).

The remaining water cannot be removed mechanically without damaging or destroying the paper web. Hence, it is dried by thermal processes. Heat energy can be introduced by conduction, convection, or radiation. All of those principals

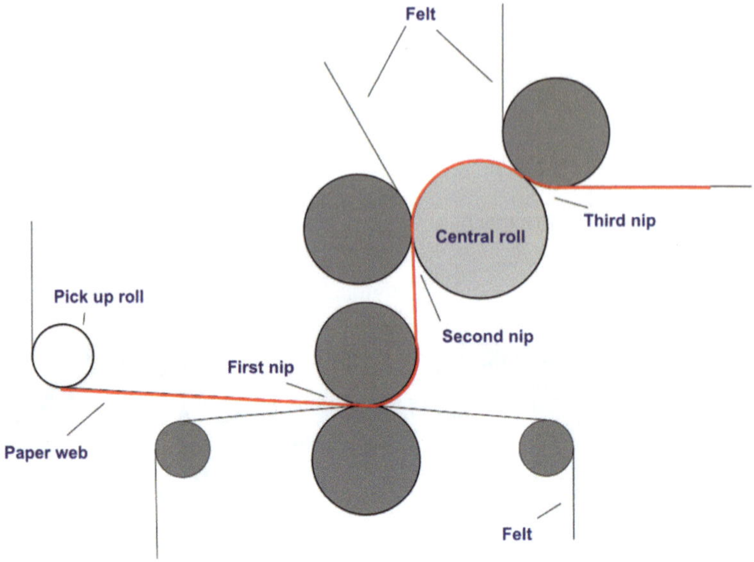

Fig. 2.11 Schematic illustration of the press section of a paper machine

Fig. 2.12 Schematic illustration of the dryer section of a paper machine

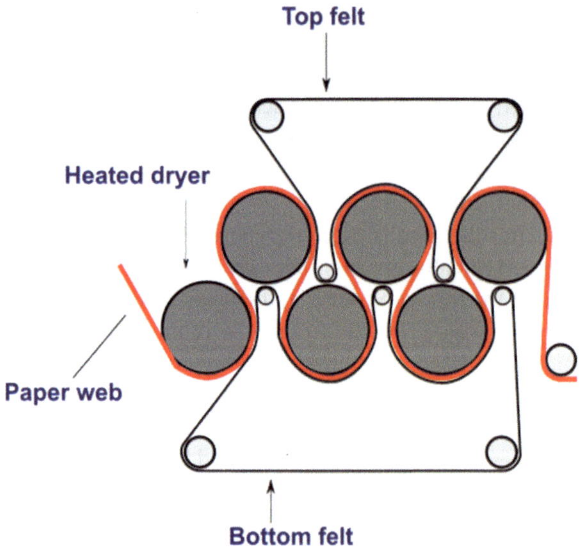

are used but most widely the paper web meandered through to alternating rolls of heated cylinders where heat is transferred to the paper web by conduction and the water in the paper web is evaporated in such a dryer section as it can be seen in Fig. 2.12.

The paper web can also be supported by felts or wires which can improve heat transfer by pressing the paper web against the dryer cylinders. These cylinders are heated by steam which is usually provided by the power plant of the paper mill. At the end of the drying section paper is dried to an equilibrium moisture of about 7–8%.

2.2 Paper Chemistry

The term "paper chemistry" often refers to chemicals being used in the pulp and paper production. Although chemicals make up only 3% of the paper as is shown in the graphical sketch with Fig. 2.13 [10], they have a very strong influence on the manufacturing processes and the properties of paper and cardboard. As such, chemicals have been used in pulp and paper production since a long time, and in particular, with respect to papermaking, in the beginning of the last century, chemical industry has increasingly been focusing on the development of synthetic paper chemicals. The latter are products, which are typically used in small quantities in paper production. Looking a bit closer, one distinguishes two types of chemical additives. So-called *process chemicals* optimize and accelerate the production process of paper, and remain in the final product only in very small quantities. *Functional chemicals*, which are added during or after sheet production, remain in the paper and serve for distinct paper properties, such as dry strength, wet strength,

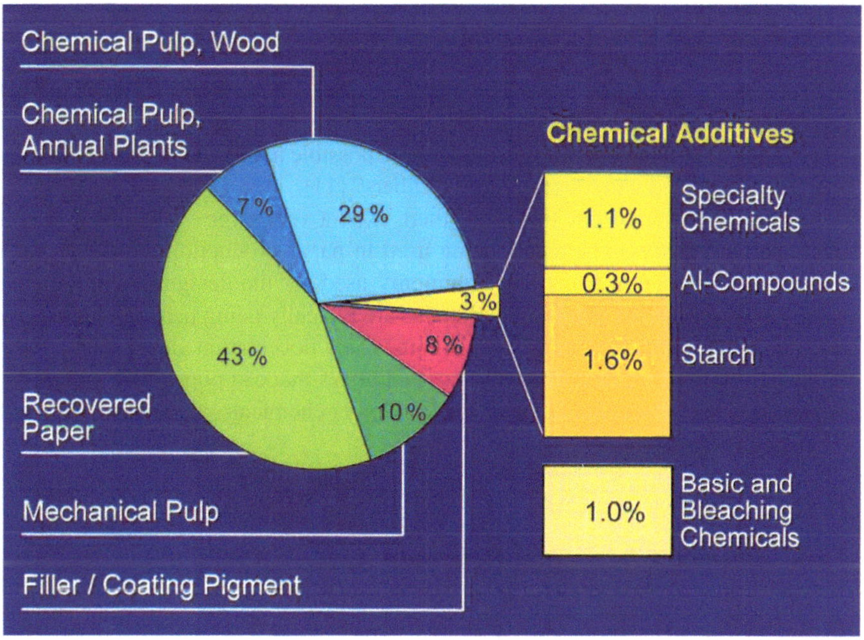

Fig. 2.13 Raw material consumption in paper and board (reprinted with permission from reference 10)

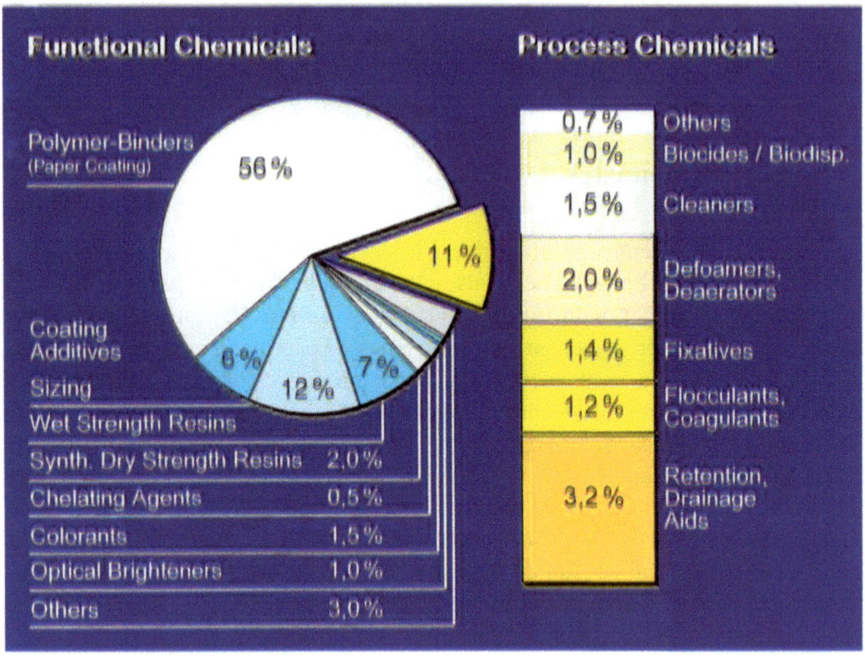

Fig. 2.14 Market shares of specialty chemicals, dry volume shares (reprinted with permission from reference 10)

or hydrophobicity of the paper and paper surface, respectively. Typical market shares of these chemicals are illustrated in Fig. 2.14. Without these two classes of chemicals, modern papermaking would not be possible at all, and paper would only be a "brown colored and loose mixture of fibers" [11].

As can be inferred from Fig. 2.14, there exist a very large variety of different functional and process chemicals being used in paper production. However, with respect to filter papers, which are commonly used for the design of microfluidic papers, only very small number of additives are typically being used, including dry and wet strength additives, as well as in some cases flocculation aids. Hence, in the following sections, we describe only those additives that can potentially impact the properties of such microfluidic papers. For all other chemicals, the reader is referred to existing textbooks [12].

2.2.1 *Process Chemicals: Retention Aids*

Because paper is being produced at very large speeds, and homogeneous fiber mat formation has to be achieved within very short times, dynamic control of the flocculation process is of utmost importance. Because cellulose fibers, including cotton linters, exhibit negative charges on the surface of the fibers which are originating from naturally oxidized OH groups of the cellulose polymers, flocculation and retention aids are indispensable for the control of homogeneous sheet formation. Common retention aids therefore consist of positively charged compounds, including inorganic compounds such as polyaluminum chloride [13, 14] or organic chemicals such as cationic polyacryl amide copolymers [15]. Such copolymers are being produced in large scale from non-charged acrylamide monomer and adding different types (and quantities) of positively charged monomers during production, i.e., polymerization. The chemical structure of monomers typically used is schematically illustrated in Fig. 2.15.

Because such retention and flocculation aids are linked to the fibers by electrostatic interactions, in principle, they can bleed off the final sheet in small quantities. Still such bleeding may interfere with any analytical process microfluidic papers are being used for. Due to the positive charges on the polymer, the macromolecules can potentially complex with negative charges that could be present with a distinct analyte, such as DNA or proteins. Knowledge of the presence and type of flocculants in a paper being used for microfluidic/diagnostic applications are therefore of importance. Yet it is not always trivial to proof for the existence of such additives in the paper, as they are being used at very small concentrations.

2.2.2 *Functional Additives: Dry and Wet Strength Additives*

Because paper fibers are mainly held together by H-bonds in the dry state, additional strengthening agents are often added that enforce the fiber-fiber joints, thereby adding to the mechanical strength in such thin paper sheets. Starch, another natural polymer, has been used as strengthening agent since a long time. Apparently, it was one of the first "added" functional additives in paper production [16]. Starch shares a common structure with cellulose, as the main building blocks both consist of glucose (Fig. 2.16). Yet, this polymer is not capable of forming fibers due to its micro-conformation, which results in highly twisted (i.e., helical) structures.

Starch is being added to paper in the wet end or after sheet formation through surface application. In both settings, the starch molecules are being dispersed in water and adsorb during drying process at the fiber-crossing points. For enhanced efficiency, starch molecules are also modified to carry positive charges prior to their use in paper production. For a detailed look into the basics of using starch as an additive, the reader is referred to existing review articles and books [17]. With respect to the use of paper in diagnostic applications, in principle, dry strength as a property may not be in focus; however, if it comes down to very thin paper sheets, dry strength additives, such as starch, ensure structural integrity of the sheets during production of microfluidic papers and during storage. Similar to retention/flocculation aids, still such additives are not firmly bound to the paper fibers and may therefore bleed off during contact with fluids, e.g., in analytical use.

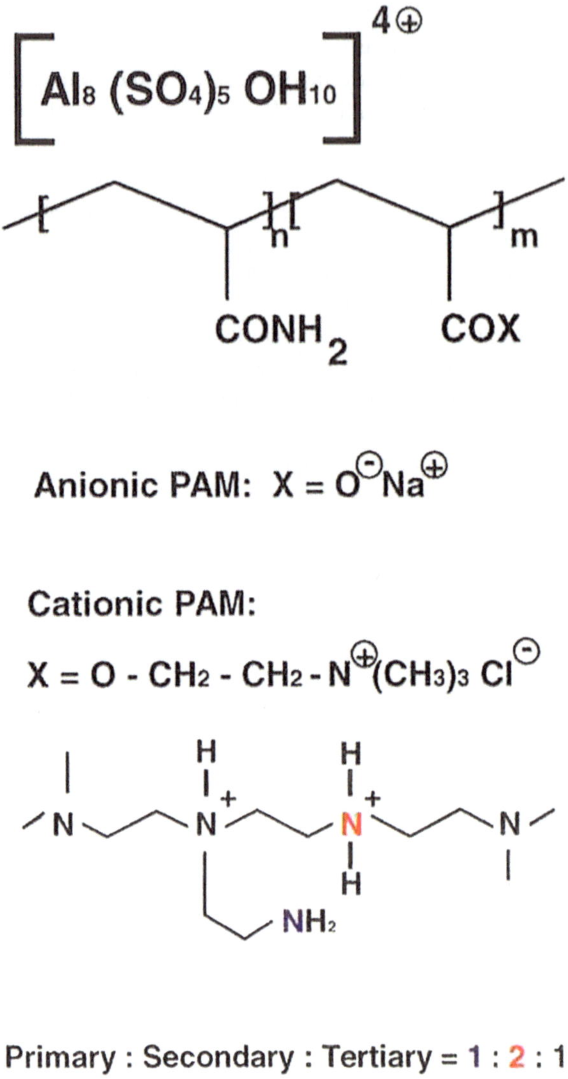

Fig. 2.15 Chemical structure of some examples of common retention aids. From top to bottom: polyaluminum chloride, polyacrylamides, and polyethylenimine (reprinted with permission from reference 10)

Finally, aside the technological application of starch, also polyacryl amide-based polymers have been used as strengthening agents.

Wet strength is an important property of paper sheets in contact with aqueous solutions. Because fibers are normally held together by H-bonds, such linkages can be easily broken by water molecules. As a consequence, structural integrity of paper sheets in the presence of water is often not trivial to retain [18]. The strength of the fiber-fiber crossings can be enhanced by either shielding such areas from contact with water ("protection") or "reinforcing" the bonding strength. Both have been shown by so-called wet-strength additives, such as melamin formaldehyde (MF)

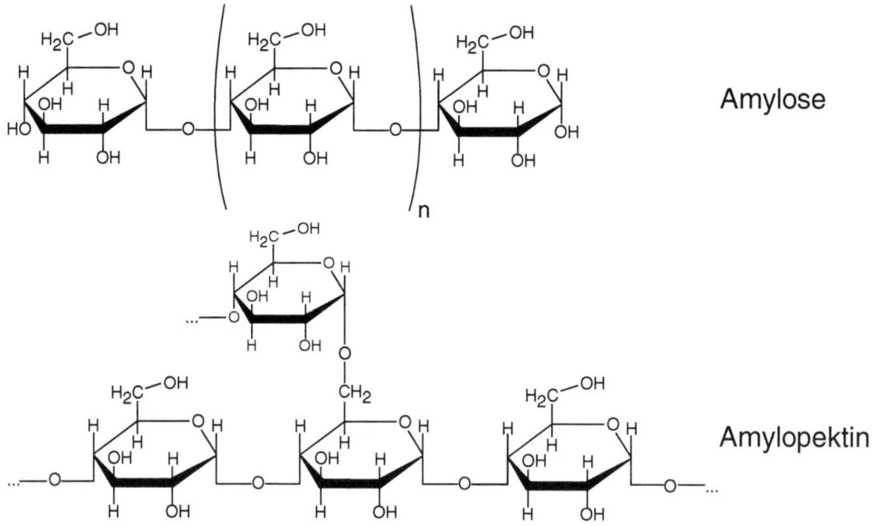

Fig. 2.16 Chemical structure of starch (reprinted with permission from reference 10)

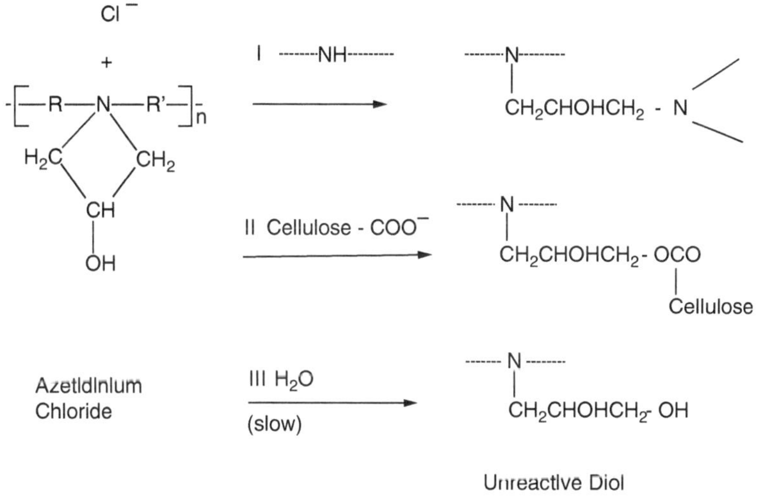

Fig. 2.17 Chemical structure of PAAE resin, and possible reactions to form network structure with cellulosic paper fibers

or polyamidoamine epichlorohydrin (PAAE) polymeric resins. These polymers are capable to from a dense polymer network surrounding the fiber-crossing points and in some cases, such as PAAE, also covalently bind to the fiber itself. The latter mechanism is shown with respect to the chemical reactions in a schematic fashion in Fig. 2.17.

Despite wet-strength agents that reinforce sheet mechanics in the wet state by thermal cross-linking mechanisms, more recently also photochemical cross-linking of fibers has been introduced [19, 20]. For any successful design of paper-based diagnostic tool it is of utmost importance to know about the presence of such wet-strength agents, because such molecules change the surface chemistry of the fibers and as such may interfere with the capillary transport of fluids as well as with any type of material brought into contact with the fibers. The latter will be discussed in more detail in the next section.

2.3 Impact of Paper Chemistry and Engineering on Paper-Based Microfluidic Devices

Although considerable progress has been made in the design of fluidic timers using paper-based substrates, to date, almost all strategies have been based on the use of simple commercially available filter papers. However, fluid transport in an engineered paper sheet does not exclusively rely on the geometric definition of the channel itself, but is strongly dependent on the intrinsic parameters of the cellulosic material such as the type of fiber used during paper fabrication, length and diameter of the fibers, swelling properties of the fibers in an aqueous environment, orientation of the paper fibers, and number of fibers per unit volume of the paper sheet (i.e., the density of the sheet). Up to date, our understanding of the influence of such parameters on the fluid transport in microfluidic papers, and how this may affect the outcome of diagnostic applications, is still in its infancy. In the following sections, we review the current state of knowledge to this respect.

2.3.1 The Role of Fiber Type

As pointed out, filter paper is often used for the development of paper-based microfluidic analytical platforms, because it consists of pure cellulose fibers typically derived from cotton linters. This pulp has only little amounts of other polysaccharides, which are often referred to as "hemicelluloses." And, more importantly, this pulp has almost no hydrophobic lignin, respectively, which can potentially interfere with analytical applications as well as impact water imbibition. Most prominent, Whatman No. 1 filter paper has been applied by many researchers for the design of microfluidic papers [21–24]. Böhm et al. used Whatman No. 1 and Roth A15 filter paper, which both consist apparently of the same type of cotton linter pulp. Both papers were compared with respect to control of a capillary-driven fluid flow [25]. Although morphological and chemical parameters were very similar with respect to chemical identity, significant different flow rates were observed. Figure 2.18 displays the outcome of this comparative study for capillary fluid flow

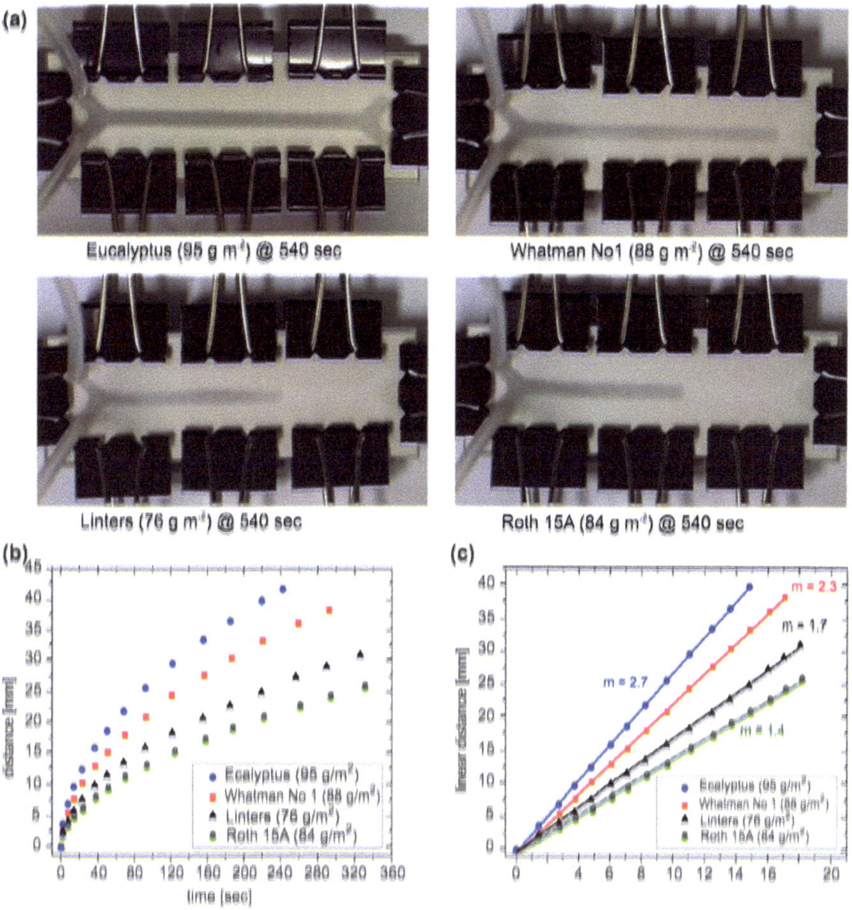

Fig. 2.18 (**a**) Comparison of the 1D position of the fluid front inside the paper-defined channels for different paper substrates at a given time interval. Photographs were taken at ~9 min after the fluid was brought into contact with the channels. (**b**) Comparison of the 1D position, x, of the fluid front driven by capillary action along the channel as a function of time for the investigated paper substrates. Individual data points were captured by video streaming the capillary transport of the fluid inside the channel. The expected relationship between distance, x, and time was observed for all paper substrates; that is, the further the distance of the fluid front from the channel inlet, the longer the time taken by the fluid front to move a certain distance inside the channel. By using different paper sources, the distance covered by the fluid front within a given time interval can be controlled over a wide range. (**c**) Comparison of the 1D position, x, of the fluid front driven by capillary action along the channel as a function of $t^{0.5}$ for the investigated paper substrates (reprinted with permission from reference 25)

in microfluidic papers which consist of different fiber types. Because both geometric structure of the sheet and surface chemistry of the fibers have a potential impact, it was concluded that by taking off-the-shelf filter paper it may be difficult to obtain reasonably reproducible results with respect to capillary flow.

Besides filter paper, the use of other paper grades, such as glossy paper (a flexible substrate made from cellulose fibers blended with an inorganic filler), has been reported [26]. Glossy paper can be considered as a good substitute for filter paper, especially when the paper surface is modified, due to its relative smooth surface. In-depth studies on the capillary-driven flow have yet not been reported.

Finally, nitrocellulose sheets have been used extensively, both in research and in technology [27–29]. Nitrocellulose membranes provide a narrow pore size distribution which results in a reproducible liquid flow within the porous network. However, the use of well-established modification techniques, such as wax printing, seems to be difficult, because penetration of hydrophobic wax into nitrocellulose membranes is slow and incomplete, if compared to filter paper. Furthermore, hydrophobic nitrocellulose membranes tend to bind biomolecules nonspecifically, including proteins [30] and DNA [31]. Finally, nitrocellulose membranes always need an additional support for structural integrity, such as thin plastic foils, because of the limited chemical and thermal stability of amorphous membrane. Hence, in comparison to paper, nitrocellulose may be more defined with respect to geometric structure; however, it is not simple to define biochemical functions and hydrophobic barriers, and thermal as well as chemical stability is an issue with such membranes.

2.3.2 *Porosity, and Fiber Pretreatment*

Customizable, lab-engineered paper sheets can allow for an accurate control of the constitution of the paper sheet itself and potential side effects caused by chemical additives or fiber pretreatments may be suppressed. One parameter, which can be adjusted by different means within the production process or by fiber pretreatment, is the porosity of the sheets. This porosity can be adjusted either by diluting a given fiber type in suspension prior to the sheet formation or by fractionation of the fibers and preparation of sheets with defined fiber-length fractions.

As illustrated in Fig. 2.19, the capillary flow inside chemically identical, tailor-made paper sheets can be controlled, by varying the fiber density (i.e., the porosity) inside the lab sheet. A similar effect can be observed, if porosity is adjusted by fiber fractionation.

Porosity of the sheets may also be tuned by fiber pretreatment. Fibers can be refined to a distinct Schopper-Riegler value, as described in Sect. 2.1. If paper sheets are made from such beaten fibers, porosity of the paper varies over a wide range. Figure 2.20 shows the capillary rise, i.e., the speed of a liquid front in cotton linter paper versus the refining degree in Schopper-Riegler. With increasing Schopper-Riegler value of the corresponding pulp, the porosity of the papers decreases. As expected the fluid transport becomes slower due to lower capillary forces, yet it still obeys the well-known Lucas-Washburn equation [32].

If only fiber fractions of distinct lengths are being used for paper manufacture, paper made of the long fibers has intrinsically larger pores than paper made from short fibers. As a consequence, capillary fluid transport in such papers is affected as well, and those papers having the largest pores exhibit fastest capillary rise as

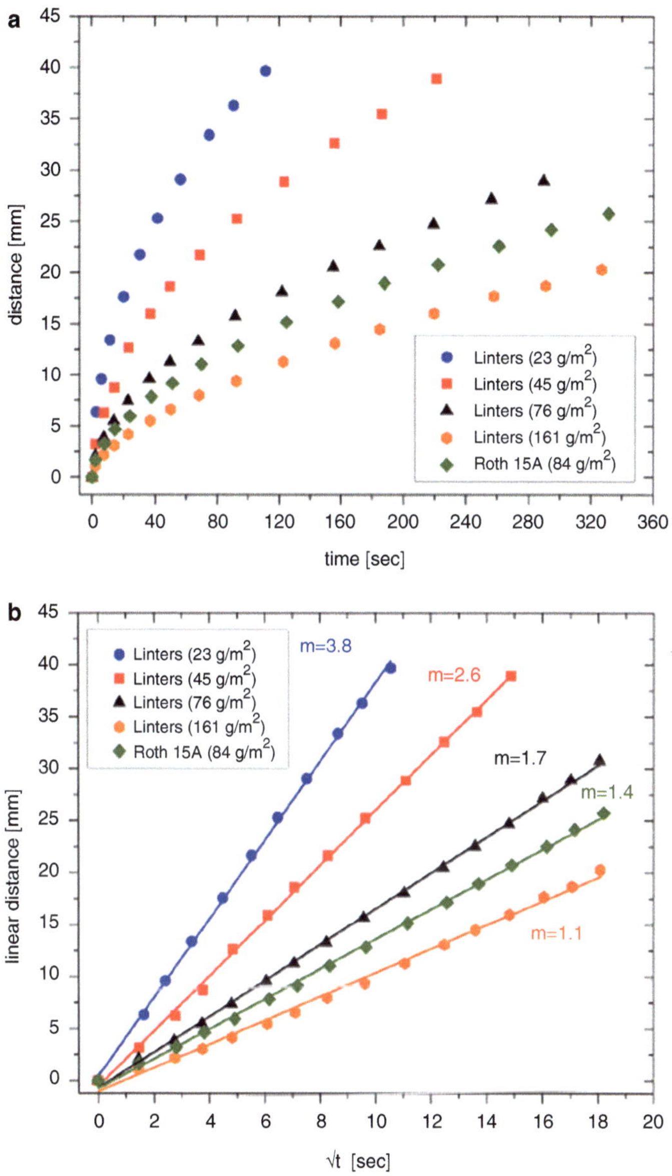

Fig. 2.19 (**a**) Comparison of the position, x, of the fluid front driven by capillary action along the channel as a function of $t^{0.5}$ for various cotton linter-based paper substrates with different fiber densities, pore sizes, and free pore volumes, and model filter paper (Roth 15A). Individual data points were captured by video streaming the capillary transport of the fluid inside the channel. A change in the fiber density of the paper substrates (effectively translating to a change in pore size, porosity, and free pore volume of the paper sheets) leads to a significant change in the distance covered by the fluid front within a given time interval. (**b**) Comparison of the position, x, of the fluid front driven by capillary action along the channel as a function of $t^{0.5}$ for various cotton linter-based paper substrates with different fiber densities, pore sizes, and free pore volumes, and model filter paper (Roth 15A) (reprinted with permission from reference [25])

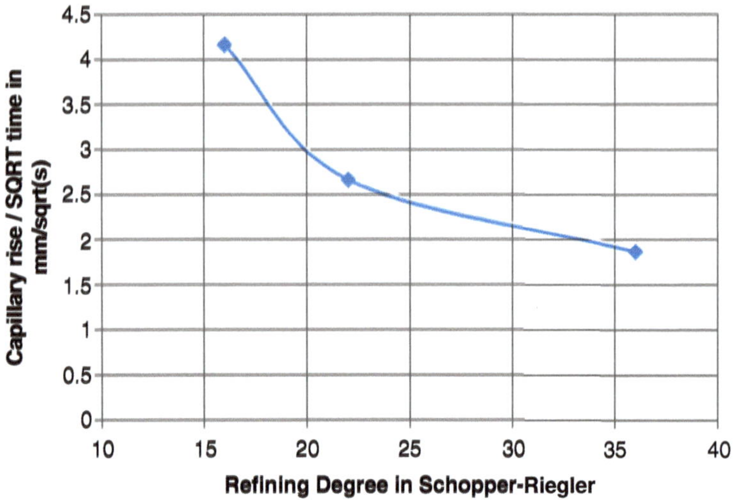

Fig. 2.20 Capillary rise in cotton linter paper for increased refining (increased Schopper-Riegler value)

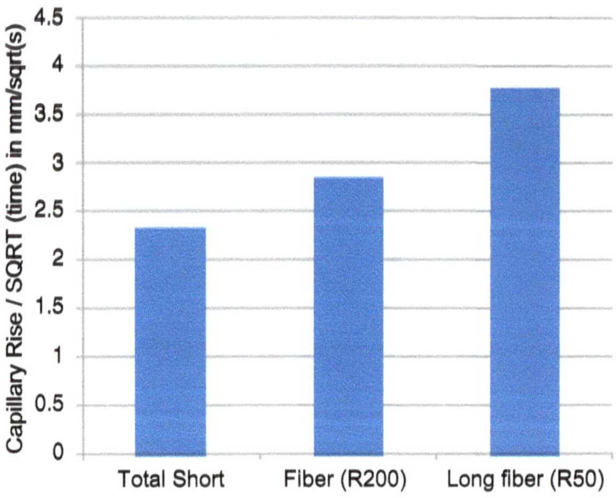

Fig. 2.21 Capillary rise in cotton linter paper for papers made from fibers of different lengths

shown in Fig. 2.21. Short fibers have smaller pores and the mixture of short and long fibers in this case results in even smaller pores and a slower capillary rise in the corresponding paper.

The examples shown in this chapter demonstrate that the transport of fluids in paper can be tailored by adjusting the pore size through either fractionation or refining of the fibers. Another parameter that has yet not been in focus concerns the surface chemistry and swelling of the fibers, which adds a further possibility for fine-tuning of capillary flow.

2.3.3 *Paper Chemical Additives*

As has been described in brief in Sect. 2.2, with paper manufacture a number of chemical additives are involved, of which some of them remain in the paper after production. With respect to paper-based diagnostic devices, polymeric retention aids as well as wet-strength-enforcing additives need to be considered. In a very recent study, Garnier and coworker have shown that both a cationic polyacryl amide (CPAM) used as retention aid and a PAAE as wet-strength resin do influence the performance of a diagnostic paper device. In first model reactions it was shown that for example red blood cells firmly bind to CPAM, thereby affecting the outcome of blood-typing experiments being performed with these papers [33].

2.4 Summary and Outlook

What do we know about paper? Does it matter whether we take an ordinary filter paper *of the shelf* or shall it be a more sophisticated and advanced paper substrate instead? What is the role of the chemical and the morphological constitution of the paper material in microfluidic and diagnostic applications? Does the type of lignocellulosic origin of the fibers, thickness, pretreatment, and porosity of the paper matter, and how? What is the role of chemical additives typically present with papers? This small book chapter, of course, cannot give all the answers to these highly interesting questions. In fact, to most of these questions our current knowledge is still very poor! Yet it provides the reader with some background information on the process of how paper sheets are produced to date, and that paper is a very complex material, both from a morphological point of view and with respect to chemical constitution. Finally, it reviews some recently published work on fundamentally understanding the role of paper manufacture and paper chemistry on the performance of paper-based microfluidic (diagnostic) devices.

However, at very early stages, it is evident that a tailored paper engineering is an important tool to fine-tune properties which guide any successful development of paper-based materials in diagnostic applications in the future. Once we understand how fiber origin, pretreatment, orientation, and surface chemistry affect fluid transport processes as well as interactions of any type of material (transported) in the pores, we may be capable to advance the current status of paper-based diagnostic devices and we will be capable to get an important step forward.

References

1. Holik H (2013) Handbook of paper and board. Wiley-VCH Verlag, Weinheim
2. CEPI key statistics (2016) www.cepi.org
3. www.marketresearchstore.com/report/global-specialty-paper-market-outlook-33404

4. Sixta H (ed) (2006) Handbook of pulp. Wiley-VCH Verlag, Weinheim
5. Gullichsen J, Paulapuro H (2000) Papermaking science and technology, chemical pulping, book 6B. Fapet Oy, Jyväskylä Gummerus Printing
6. Gullichsen J, Paulapuro H (1999) Papermaking science and technology, mechanical pulping, book 5. Fapet Oy, Jyväskylä Gummerus Printing
7. Gullichsen J, Paulapuro H (2000) Papermaking science and technology, recycled fiber and deinking, book 7. Fapet Oy, Jyväskylä Gummerus Printing
8. www.voith.com
9. Niskanen K (2011) Mechanics of paper products. De Gruyter, Berlin
10. Kleemann S et al (2008) Chemical additives for the production of pulp & paper. Deutscher Fachverlag, Frankfurt am Main
11. Thummer R, Belle J (2008) Chemical additives for the production of pulp & paper. Deutscher Fachverlag, Frankfurt am Main
12. Alen R (2007) Papermaking chemistry, book 4. Fapet Oy, Jyväskylä., Gummerus Printing
13. Auhorn WJ, Niemelä K (2006) Process chemicals for the production of chemical pulp. Wochenbl Pap 134(22):1302
14. Kuenzel U, Le PC (2008) In: Kleemann S et al (eds) Chemical additives for the production of pulp & paper. Deutscher Fachverlag, Frankfurt am Main
15. Pelzer R (2008) In: Kleemann S et al (eds) Chemical additives for the production of pulp & paper. Deutscher Fachverlag, Frankfurt am Main
16. Auhorn WJ (2008) In: Kleemann S et al (eds) Chemical additives for the production of pulp & paper. Deutscher Fachverlag, Frankfurt am Main
17. Hubbe MA et al (2007) What happens to cellulosic fibers during papermaking and recycling? A review. Bioresources 2(4):739–788
18. Belle J, Odermatt J (2016) Initial wet web strength of paper. Cellulose 23:2249–2272
19. Jocher M et al (2015) Enhancing the wet strength of lignocellulosic fibrous networks using photo-crosslinkable polymers. Cellulose 22(1):581–591
20. Janko M et al (2015) Cross-linking cellulosic fibers with photoreactive polymers: visualization with confocal Raman and fluorescence microscopy. Biomacromolecules 16:2179–2187
21. Böhm A et al (2013) Photo-attaching functional polymers to cellulose fibers for the design of chemically modified paper. Cellulose 20(1):467–483
22. Liana DD et al (2012) Recent advances in paper-based sensors. Sensors 12:11505
23. Dungchai W et al (2009) Electrochemical detection for paper-based microfluidics. Anal Chem 81:5821–5826
24. Wang J et al (2014) Hydrophobic sol-gel channel patterning strategies for paper-based microfluidics. Lab Chip 14:691–695
25. Böhm A et al (2014) Engineering microfluidic papers: effect of fiber source and paper sheet properties on capillary-driven fluid flow. Microfluid Nanofluid 16:789–799
26. Arena A et al (2010) Flexible ethanol sensors on glossy paper substrates operating at room temperature. Sens Actuators B Chem 145:488–494
27. Lu Y et al (2010) Fabrication and characterization of paper-based microfluidics prepared in nitrocellulose membrane by wax printing. Anal Chem 82:329
28. Fu E et al (2010) Controlled reagent transport in disposable 2D paper networks. Lab Chip 10:918
29. Fu E et al (2011) Transport in two-dimensional paper networks. Microfluid Nanofluid 10:29
30. Fenton EM et al (2009) Multiplex lateral-flow test strips fabricated by two-dimensional shaping. ACS Appl Mater Interfaces 1:124
31. Cretich M et al (2010) Coating of nitrocellulose for colorimetric DNA microarrays. Anal Biochem 397:84
32. Washburn EW (1921) The dynamics of capillary flow. Phys Rev 17:273
33. McLiesh H et al (2015) Effect of cationic polyelectrolytes on the performance of paper diagnostics for blood typing. Coll Surf B Biointerfaces 133:189–197

Chapter 3
Critical Components and Innovations in Paper-Based Analytical Devices

Robert B. Channon, Monpichar Srisa-Art, Katherine Boehle, and Charles Henry

Despite a long history as an analytical tool, paper has come into significant academic and industrial focus over the last ~10 years. This resurgence coincided with the seminal work by Whitesides group in 2007 on photoresist patterning of paper to create well-defined microfluidic channels [1]. The exponential growth can in large part be linked to key innovations, which have taken paper-based analytical devices from a niche apparatus in a handful analytical labs to a powerful tool used across a variety of disciplines around the world. Here, we outline some of the key advances and their impact.

3.1 Flow Manipulation Tools

Paper-based analytical devices can broadly be segregated into three distant factions based on their design, namely spot tests, distance-based devices, and microfluidic devices. Spot tests have been popular historically due to their simplicity and have been deployed across a wide range of fields from environmental monitoring to disease diagnosis. More specific information on spot tests, stationary PADs, is covered elsewhere in this book, as well as in several literature reviews [2, 3]. PADs employing fluid flow are commonly known as microfluidic paper-based analytical devices (μPADs). The recent general movement in academia and industry from stationary PADs to μPADs can be understood in terms of several important advantages of μPADs over their stationary counterparts:

R. B. Channon · K. Boehle · C. Henry (✉)
Department of Chemistry, Colorado State University, Fort Collins, CO, USA
e-mail: Chuck.Henry@ColoState.edu

M. Srisa-Art
Department of Chemistry, Chulalongkorn University, Bangkok, Thailand

K. J. Land (ed.), *Paper-based Diagnostics*,
https://doi.org/10.1007/978-3-319-96870-4_3

- Sample transport facilitates online sequential reactions and preconcentration of a target analyte that are challenging to implement in a spot test. These analytical tools are covered in more detail later in this chapter.
- Addition of flow to an analyte sample often results in increased sensitivities and lower detection limits for many detection methods compared to stationary detection [4].
- For colorimetric reactions (reaction with an analyte resulting in a color change), reactants can be deposited along a channel to react with a target analyte during fluid transport, resulting in a set distance of color generation, corresponding to the analyte concentration. Distance-based detection is a popular method for colorimetric PADs, particularly in point-of-care settings due to the simplicity of readout [5].

3.1.1 *Wicking Speed*

Fluid flow in paper has been around as early as 1956 through lateral flow tests [6], for example the common pregnancy test. Crucially, fluid flow is generated through capillary action in the paper fibers. This circumvents the need for external pumps making μPADs cheap, small, and simple to use. Over time, viscous resistance of the fluid acts counter to the capillary action, slowing down the fluid velocity. The distance traversed down the channel, $l(t)$ (m), at time t (s), is given as the Lucas-Washburn equation [7]:

$$l(t) = \sqrt{\frac{\gamma r t \cos\theta}{2\mu}} \tag{3.1}$$

where γ is the interfacial tension (N m^{-1}), r is the mean capillary radius (m), θ is the fluid contact angle on the paper, and μ is the fluid viscosity (N s m^{-1}). This equation assumes (i) a constant cross section of cylindrical straight pores, (ii) that gravitational effects are negligible (i.e., horizontal flow), (iii) that the paper is chemically homogenous, and iv) an unlimited inlet reservoir volume.

μPADs can be designed to generate a constant flow rate, for example through an increasing channel geometry [8]. In this case, the steady-state flow rate is given by Darcy's law [9]:

$$Q = \frac{-\kappa A \Delta P}{\mu L} \tag{3.2}$$

where Q is the volumetric flow rate (m s^{-1}), κ is the paper permeability (m^2), A is the cross-sectional area of the paper normal to flow (m^2), and ΔP is the pressure difference (N m^{-2}) over length L down the channel (m). This equation assumes (i) negligible kinetic energy, (ii) a circular fiber cross section, and (iii) straight

Table 3.1 Methods to control the flow rate in μPADs, organized in terms of ease of implementation (from low to high)

Property	Method of control	Effect on flow rate
Paper type	Vary mean capillary radii	Larger pore sizes increase the flow rate (Eq. 3.1)
Encapsulation	Open vs. sealed channels	Sealing inhibits evaporation leading to longer flow distances, but can reduce the flow rate if sealing tight (e.g., lamination or pressing)
Paper modification	Printed wax lines or micro-channels cut into the paper	Small wax-printed lines can reduce capillary action for slower flow, micro-channels can be cut into paper to speed up flow, ionogel/sugar deposition can be used to slow down flow
Geometry	Width of channel, e.g., distance between wax barriers	Wax and other common materials have lower contact angles, e.g., thinner channels have less resistance and faster flow rates; conversely, a 270 "fan" geometry generates steady-state flow
Temperature	Change temperature in lab	Higher temperatures result in reduced viscosity and faster flow, temperature-sensitive hydrogels have also been used to control the flow rate
Volume	Vary volumes of sample added to device	Sample flow requires an excess of sample in the inlet, so flow rates can be slowed/stopped through small inlet volumes
Orientation	Change angle of device, e.g., vertical or horizontal	Gravitational forces can inhibit or enhance flow rates based on flow direction, e.g., in vertically oriented devices, the flow can cease when the capillary and gravitational forces balance
Number of paper layers	Multiple paper layers or hollow channels	Creation of a hollow channel via a gap above the paper or multiple layers provides significant enhancements in flow rate
Head pressure	Inlet wells raised above the channel	Use of inlet wells drives the flow through head pressure, making use of gravity

capillaries. These equations are by no means exhaustive perfect models for fluid flow in all paper devices, but serve as valuable approximations of real systems. Various other analytical and computation models have been proposed to account for additional variables, such as evaporation (humidity), different fluid barriers (e.g., wax), and multilayer devices (e.g., flow between two layers of paper) [10, 11], though none have yet achieved widespread adoption by the field. Control over fluid velocities and volumetric flow rates is critical to proper use of μPADs. For example, enzymatic reactions and immunoassays in paper require slow flow rates, but measurement of quasi-stable reagents or analytes at trace concentrations requires high flow rates. Table 3.1 lists some tunable properties of μPADs and their effect on flow rates.

Generation of steady state or constant flow rates is critical for many analytical measurements. For example, electrochemical redox currents are proportional to the cubed root of the flow rate (for rectangular channels) [12], and thus small changes in

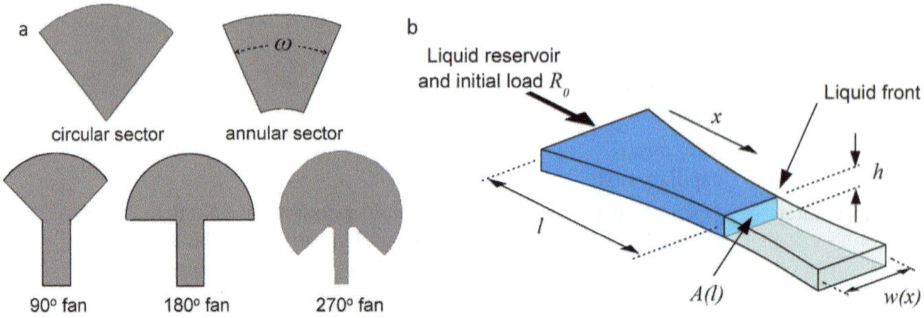

Fig. 3.1 Two approaches for quasi steady-state flow in μPADs. (**a**) Different downstream channel geometries to generate steady-state flow, (**b**) model of fluid flow in expanding channel width μPADs, adapted with permission from references [8, 157], respectively

flow rate during an experiment can have significant effects on the analytical signals. Quasi steady-state flow can be achieved through an increase in cross-sectional area as shown in Fig. 3.1.

Here, an increase in channel width/pore volume results in steady-state flow rate as the fan/pumping region imbibes fluid with a constant force. Flow rates in these systems are typically around 0.020 cm s^{-1} or 7.1×10^{-6} mL min^{-1}.

A sizable section of recent research on μPADs is focused around methods to increase velocities and volumetric flow rates. Advancements range from simple geometric changes to the addition of more complex analytical tools such as syringe pumps. One elegant solution was proposed by Crooks et al. [13] based on pressure-driven flow in a hollow channel as shown in Fig. 3.2a. These hollow-channel PADs (HC-PAD) were applied for electrochemical detection under flow rates of 1.2 cm s^{-1} or 0.21 mL min^{-1}, through applied pressure from an inlet well coupled with the hollow channel. This approach has been expanded by Henry and coauthors [4] to achieve flow rates of up to 1.56 cm s^{-1} or 1.65 mL min^{-1} in multilayered μPADs, through precise control of the channel height/gap between the paper layers (up to 390 μm, Fig. 3.2b).

3.1.2 Switches and Valves

The controlled manipulation of fluid flow is a critical tool in μPADs. This is particularly true for multistep online reactions where different flow rates/reaction times are required for different assay steps. While simple to implement through manual control in classical analytical assays (e.g., through operator control), this is more challenging to implement in μPADs, particularly without manual operator control. The ideal μPAD would allow direct control to stop/start flow (at different flow rates), for a large number of online assay steps. Various research groups around the world have taken significant steps towards this goal.

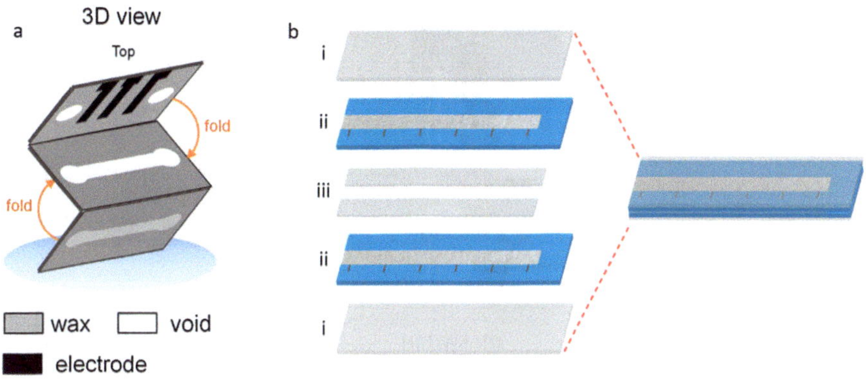

Fig. 3.2 Rapid-flow (**a**) hollow channel and (**b**) multilayered μPADs (i—packing tape, ii—wax-modified Whatman 1 qualitative paper, and iii—double-sided tape) adapted with permission from references [4, 13], respectively

Manual control methods use operator involvement to control the flow rates, allowing for precise fluid control. Mechanical switches are the most common approach for fluid control as they do not require expensive external equipment, making them well suited for use in point-of-care environments. Pull slips employ user-pulling of a strip which can then open up/close a channel to start/stop fluid flow. These have been used for a variety of applications such as paper-based enzyme-linked immunosorbent assays (ELISAs) [14]. An example device where sliding of the sample layer initiates flow is shown later in this chapter in Fig. 3.5 (Sect. 3.2.2). Similarly, tabs or folding has been demonstrated as valves, where pushing down a tab/folding a device connects paper channels to start flow. A simple tab/valve design is shown in Fig. 3.3a, where reagent flow into a reactor is initiated by pressing down a valve [15]. Folding devices to initiate flow/reagents steps has also been previously demonstrated. For example, Fig. 3.3b shows an origami-type folding device for point-of-care nucleic acid extraction. The layers L1–L4 fold above and below layer L0, and the device functionalities represent *1* DNA filter, *2* waste absorption pad with Mylar backing, *3* sample loading cup, *4* lysis/wash buffer storage and rehydration pad, *5* buffer transport channel, and *6* contact stack [16]. Manual operator switches can also employ time or volume controls to start, stop, or delay sample flow. For example, Fig. 3.3c, d shows an on switch, where fluid flow down a "timing channel" swells an actuator facilitating delayed sample fluid flow, with different timing channel lengths resulting in different delays.

Nonmechanical operator-controlled methods are less common, and usually employ temperature, electric, or magnetic switches and valves. Temperature can be used to vary the hydrophobicity of paper modified with a surfactant. For example, hexadecyltrimethylammonium bromide (CTMAB)-modified paper is hydrophobic at low temperatures (inhibiting fluid flow), but becomes hydrophilic at higher

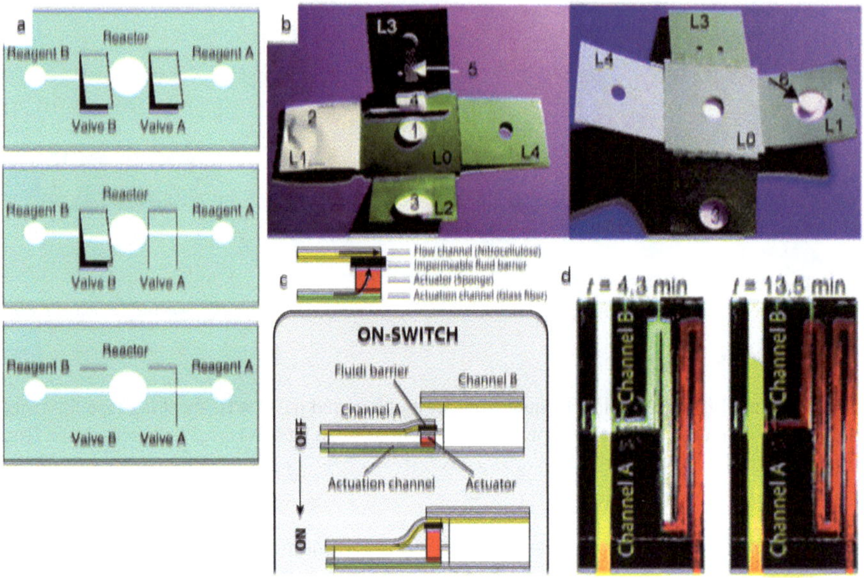

Fig. 3.3 Operator-controlled mechanical switches, (**a**) a simple push-tab valve design for introduction of reagents to a reactor [15], (**b**) a microfluidic origami device for nucleic acid extraction (left: front side, right: backside of device) [16], (**c**) a time-metered on switch for the controlled delay of sample flow [158], adapted with permission from references [15, 16, 158], respectively, (**d**) Time-lapse images of an on-switch with a serpentine actuation channel to provide timed valve actuation

temperatures (enabling fluid flow) [17]. Similarly, dielectrics can be employed to change the hydrophobicity of a paper channel, based on application of a voltage to hydrophobic/hydrophilic electrodes to stop/start fluid flow [18]. Magnetic switches have been developed to delay sample flow for ~30 min, based on an ionic resistor triggering a electromagnetic which opens/closes a paper cantilever valve, starting/stopping the flow [19].

Automated methods are intended to control flow without operator input and are typically designed for specific volumes, flow rates, and timings towards specific applications. A simple and elegant example of this involves cutting micro-channels into paper channels using a knife blade [20]. When the channels are parallel to the channel, flow rates are enhanced by ~60% compared to untreated devices. Moreover, flow rates can be slowed by ~45% through micro-channels carved perpendicular to the direction of flow. Changes in geometry/cross-sectional area can also be used to in situ generate fluid delays. For example, placement of an absorbent pad/shunt can generate delays of up to 20 min [21]. Finally, paper modification can be used to in situ generate a delay in the fluid flow. One common example of this is through the deposition of sugars, where a higher concentration of sugar results in a slower flow rate/longer delay [22].

3.2 Sample Loading

Sample addition is a critical aspect of PADs which often receives little consideration despite its importance. Aside from sample pretreatments which will be covered later in this chapter (Section 3.3), there are a variety of approaches towards sample loading, each with its own advantages and drawbacks. Crucially, the approach must be geared towards the circumstances; pipetting a set volume of analyte at a set time to a device by a trained technician is ideal for centralized lab testing. Conversely, much recent research in the paper-based field is focused towards the swift, robust, controlled, and sequential injection of multiple reagents, at set times, in controlled conditions, within a resource-limited setting, without operator involvement and at low cost.

3.2.1 *Single Delivery*

Sample delivery to a PAD can be carried out through external approaches (e.g., pipetting), or online approaches (e.g., blister packs). In both cases the specific volume and timing of delivery are the critical factors. Some common manual approaches employing operator instigation are detailed in Table 3.2.

To achieve device deployment in resource-limited settings, many new innovations seek to apply operator-free loading of controlled sample volumes. The sample

Table 3.2 Manual approaches for single-reagent delivery in PADs

Approach	Common application	Advantages/drawbacks
Manual pipetting	Reagent delivery in centralized/research laboratories	Ideally suited for research and control of volume and injection time, but requires trained technician
Disposable pipettes	Environmental sampling, e.g., wastewater sampling	Simple, cheap for in-field testing, but non-disposable and inherent variability with amateur operator
Finger pricks	Blood sampling	Can be used by nonspecialists at the point of care, but transfers variable and unknown volume
Microneedles [23]	Blood sampling	More accurate volume than finger pricks and integrated with PAD, but more complex device
Slipchip [24]	Parallel processing/addition of reagents to a series of wells/devices	High-throughput/multiplexed sensing facilitated, but requires preloading of reservoirs with pipette

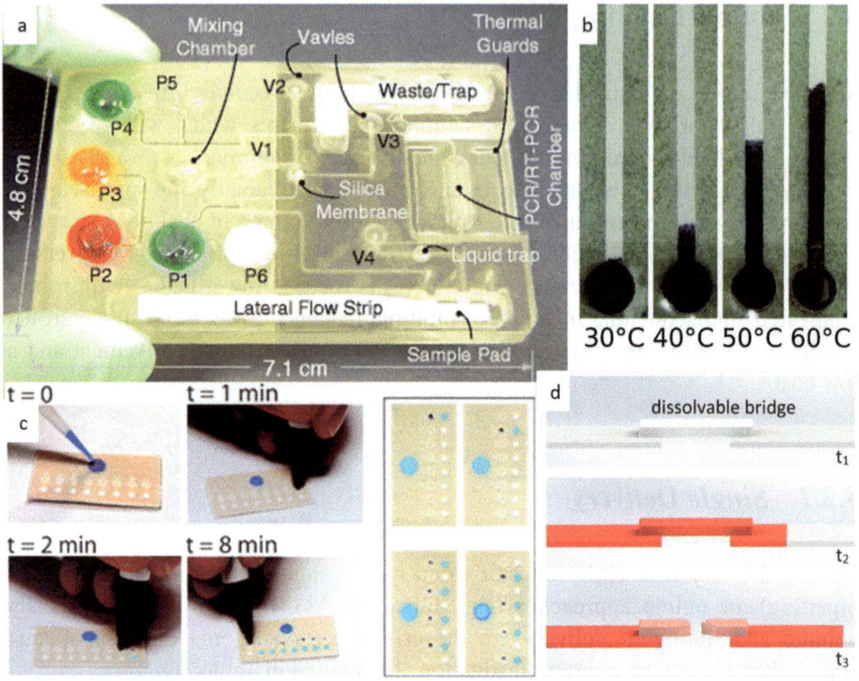

Fig. 3.4 On-device sample addition methods, (**a**) blister packs (p1–p6) for sample injection to a lateral flow assay [25], (**b**) temperature-mediated sample release from a hydrogel [27], (**c**) pressing "on" buttons to initiate sample addition to different regions of a μPAD [26], (**d**) dissolvable sugar bridges for sample volume control [28], reproduced with permission from references [25–28], respectively

volume is a particularly critical variable, as it affects the mass/particle loading from a real sample and thus must be strictly controlled for analytical testing. Blister packs are an excellent tool for storing a set reagent volume on a device, which can be released through pressing on the pouch by an operator, or through closing of a cassette. An example device is shown in Fig. 3.4a, where reagents are injected from the blister packs through device-controlled plungers which are housed on the cassette, resulting in controlled sample injection towards a lateral flow assay for nucleic acid detection [25]. A similar concept has been demonstrated for sample addition to different regions of a device, through pressing of spots with a pen to close channels and initiate fluid flow, as shown in Fig. 3.4c [26]. Hydrogels can be used to store reagents and release them from external stimuli, such as a temperature change. Figure 3.4b demonstrates this concept with linear solute release from a composite *N*-isopropylacrylamide/acrylamide hydrogel on increasing temperatures [27]. One interesting solution to controlling sample volumes is through dissolvable bridges as shown in Fig. 3.4d. Here, the sugar bridge can be tuned to deliver different volumes based on sugar concentration and geometry [28].

3.2.2 Sequential Delivery

Sequential delivery or injection is naturally more complex than single delivery and has come into recent focus by the field over the last ~5 years. The ability to add different reagents of different volumes and at controlled times and flow rates is critical for the development of μPADs, given that many analytical assays involve multiple steps and controlled times/reaction conditions. While deposition of reagents onto a μPAD for later rehydration can facilitate sequential reactions, this format is inflexible and only applicable to reagents with appropriate solubility for deposition that will be stable when dried onto paper.

One of the simplest approaches towards sequential reagent delivery is to employ varied path length channels, where longer inlet paths take longer to transverse allowing for timed sample addition. One example of this is shown in Fig. 3.5a; for an ELISA an antibody (c) and substrate (d) are added sequentially towards downstream test (b) and control (a) zones [29]. One drawback of this approach is that longer channel lengths result in greater sample loss and thus the majority of sequential injection approaches employ alternate methods [30]. As previously described, sugar deposition can be used to delay fluid flow, and this has been demonstrated for sequential injection based on the concentration of deposited sugar in different channels of a folding cassette as shown in Fig. 3.5b [22]. Finally,

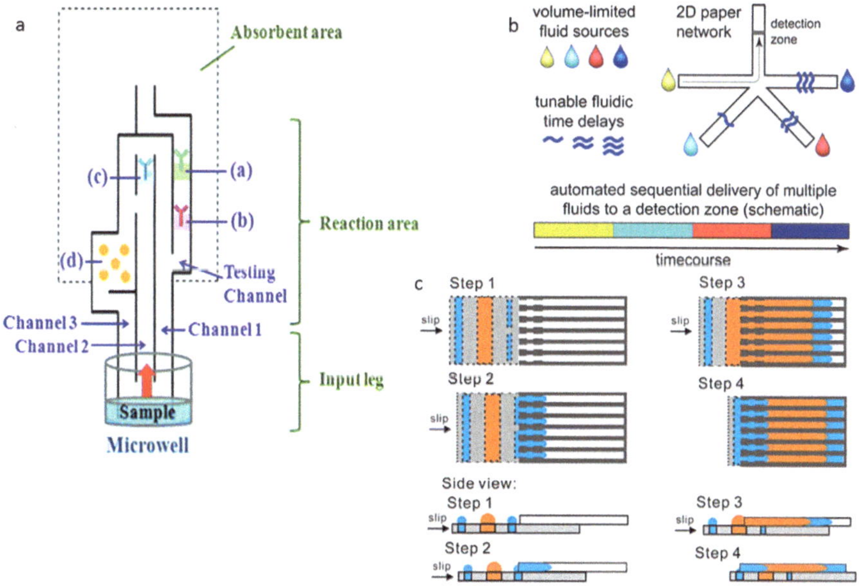

Fig. 3.5 Simple sequential delivery paper devices, (**a**) sandwich ELISA based on channel lengths [29], (**b**) sugar-modified channels for sequential injection from identical length channels [22], (**c**) slipPAD for sequential delivery through a pull strip [24], reproduced with permission from references [22, 24, 29], respectively

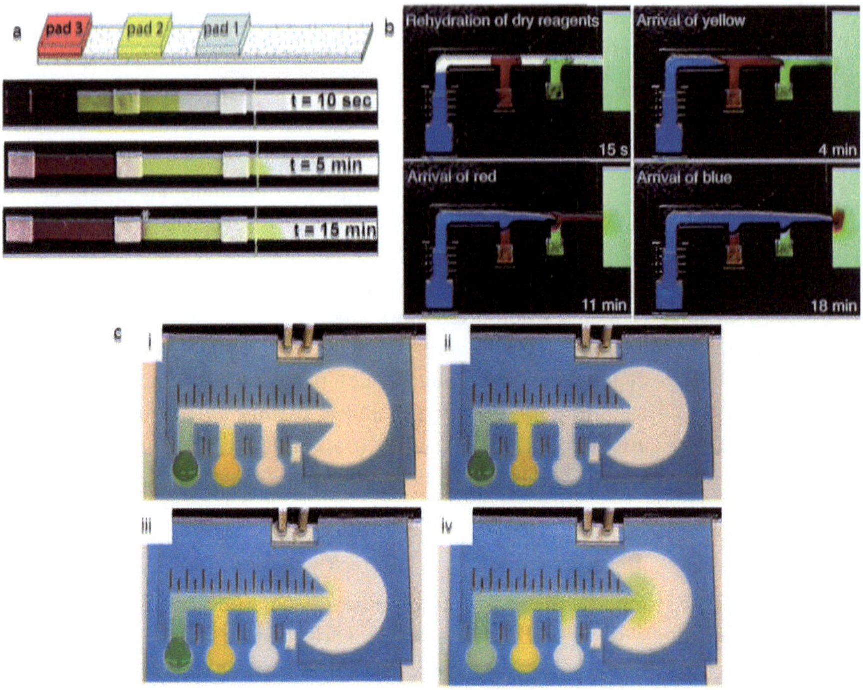

Fig. 3.6 Paper network sequential injection devices, illustrated with different dyes, (**a**) 1DPN [31], (**b**) 2DPN [32], (**c**) 3DPN [4], reproduced with permission from references [4, 31, 32], respectively

slipPADs employing pull strips can be employed for the sequential delivery of preloaded reagents, as shown in Fig. 3.5c [24].

Paper networks are a useful class of sequential injection μPADs, with some advanced functionalities over the previous devices. Example one-, two-, and three-dimensional paper networks (1DPN, 2DPN, 3DPN) are shown in Fig. 3.6. 1DPNs feature pads or wells to hold the different reagents which then sequentially release along a straight paper channel (Fig. 3.6a) [31]. 2DPN employ inlet legs parallel to the main paper channel which can be initiated on reagent delivery to the legs, or preloaded with reagents (sample dried onto the legs for rehydration and sequential delivery) [32]. The samples deplete sequentially from the downstream legs and these 2DPN have been tested with a variety of applications [33]. 3DPN employ the same inlet leg principles as 2DPN, but employ a large hollow channel (390 μm) between two paper layers as previously described in Fig. 3.2 (Sect. 3.1.1), resulting in significantly faster flow rates [4]. 1DPN and 2DPN feature slower flow rates than 3DPN (1 vs. 15 min), but simpler device architectures and fabrication schemes. These paper network devices are more amenable than the previous sequential injection PADs for applications with large sample volumes, quasi-stable reagents, or analytical methods where the flow rate contributes to the signal (e.g., electrochemistry).

3.3 Sample Pretreatment

When detecting a specific analyte in environmental or biological samples, many components are present in the matrix beyond the analyte of interest. While these additional components may be inert to the assay, in many cases they can cause false positives or negatives, necessitating additional pretreatment steps to remove them. Additionally, one common pitfall of PAD is higher LODs compared to traditional methods, so sample preconcentration is often required to reach the required LODs depending on the application. Herein, we discuss some of the several approaches to remove interferences from complex samples and preconcentration tools.

3.3.1 *Removing Interferences from the Sample Matrix*

When detecting specific metals in environmental samples, it is important that the detection method does not cross-react with other metallic species, causing false positives (Fig. 3.7a, b). Unfortunately, this is a common challenge faced in colorimetric detection motifs for metals. Masking agents can be used to prevent the colorimetric substrate from reacting with interference metals [2]. Masking agents

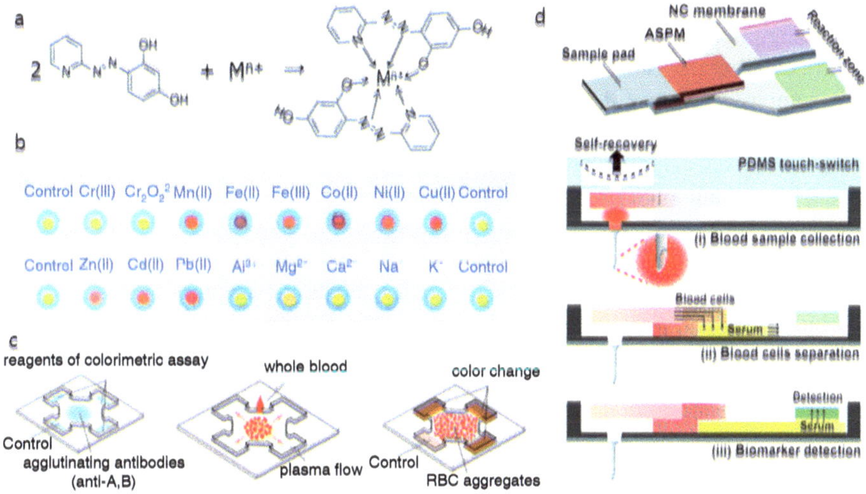

Fig. 3.7 (**a**) Complexation of 4-(2-pyridylazo)resorcinol (PAR) and metal species (M^{n+}) [36]; (**b**) paper-based spot test showing metal-PAR reactivity (orange and red color formation) for several transition, alkali, and alkaline earth metals [36]; (**c**) paper-based device that first separates red blood cells from plasma through agglutination, and then detects glucose through colorimetry [38]; (**d**) touch-activated multidiagnostic system with asymmetric polysulfone and nitrocellulose membranes for plasma separation [23], reproduced with permission from references [23, 36, 38], respectively

could be reagents that are used to adjust the pH or oxidize or reduce a particular metal species, facilitating selectivity [5, 34, 35]. Helpfully, paper can be patterned such that a single sample addition to a μPAD can be guided into several different channels, enabling users to screen for several masking agents at once [36]. Masking efficiency of a specific chemical used for interfering species is then defined by percentage change in signal between the masking intensity and the control intensity.

Biological samples are typically complex, including a variety of components in the sample beyond the biomarker or pathogen that user detects. For example, whole blood is a cocktail of immunological and molecular biomarkers that are used for disease diagnostics, making it a critical sample matrix for biomedical applications of paper-based devices. Due to the complexity of whole blood, many diagnostic motifs remove red blood cells for detection, necessitating field-ready solutions to filter and process blood. In diagnostic laboratories, centrifuges are a common tool for separating red blood cells from plasma. Paper has been used to develop a hand-powered centrifuge for in-field use based on the physics of a whirligig toy, which can reach speeds of up to 125,000 RPM [37]. Additionally, because μPADs are most commonly fabricated using filter paper, many devices have been developed where blood filtration and biomarker detection are completed in situ. Most paper-based devices that employ blood filtration filter plasma from whole blood based on aggregation of red blood cells [38, 39] or size exclusion [40]. Following filtration, the plasma is used for biochemical tests to detect specific biomarkers such as antibodies specific to blood typing or glucose (Fig. 3.7c, d) [23, 38, 39].

While paper is a valid platform for whole-blood processing and filtration, it is not as applicable to other biological samples. There are many complex sample matrices where specific biomarkers must be filtered from other components of the sample so as not to interfere with the detection assay. For these samples, users can implement immunomagnetic separation (IMS), where antibodies specific to a biomarker or pathogen are covalently attached to the surface of a magnetic bead. The immunomagnetic beads are added to a complex sample to adhere to the analyte before separating the beads from the sample using a magnet, followed by reconstituting the beads in fresh buffer. IMS has been used beyond paper-based devices for many different complex samples, including meat and dairy, as well as biological samples like whole blood and even animal feces [41]. In μPADs specifically, IMS has been demonstrated for the detection of *Escherichia coli* and *Salmonella typhimurium* in water, whole milk, and bird feces [42, 43].

3.3.2 Preconcentration

Aside from overcoming matrix effects, analyte concentrations in real biological or environmental samples can be highly scarce. Therefore, the ability to preconcentrate a sample, especially online, is a powerful tool to achieve the trace detection limits that are regularly required. For example, as well as filtering the target analyte from its matrix, IMS also facilitates preconcentration through reconstitution in a smaller

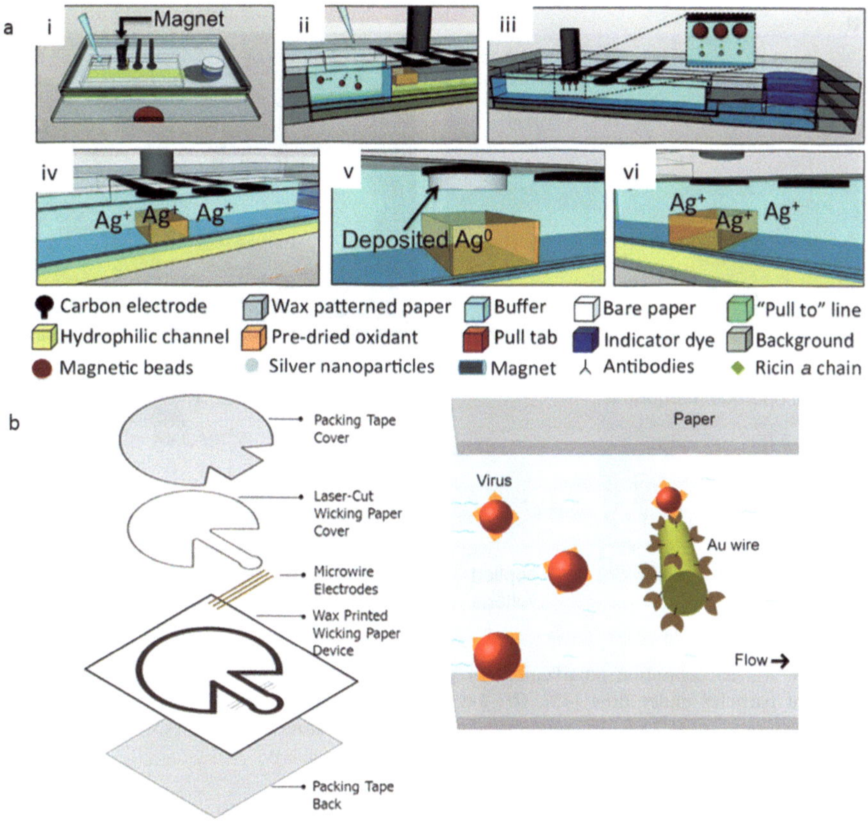

Fig. 3.8 Preconcentration through downstream capturing, (**a**) magnetic capturing of ricin/antibody/magnetic beads [159], (**b**) capturing of virus particles on antibody-modified microwire electrodes in a multilayered μPAD [44, 160], reproduced with permission from references [44, 159, 160], respectively

volume of buffer than the initial sample volume. One of the simplest approaches for preconcentration is to capture the target of interest onto a sensor downstream during flow. This can be carried out through binding of a target to magnetic particles, for preconcentration downstream as shown in Fig. 3.8a. Similarly, virus particles have been captured on antibody-modified electrodes placed downstream in a hollow-channel multilayer μPAD as shown in Fig. 3.8b [44].

Ion concentration polarization (ICP) is a powerful technique for the preconcentration of charged species. Here, application of an electric field separates charged particles from the rest of a sample to facilitate preconcentration. For example, charged fluorescent dyes can be separated from the solute into different fluidic channels downstream, on application of charge from Nafion electrodes, as shown in Fig. 3.9a [45]. A similar approach can be used for the separation and detection of ds- and ssDNA, using an electric field for electrophoretic migration and

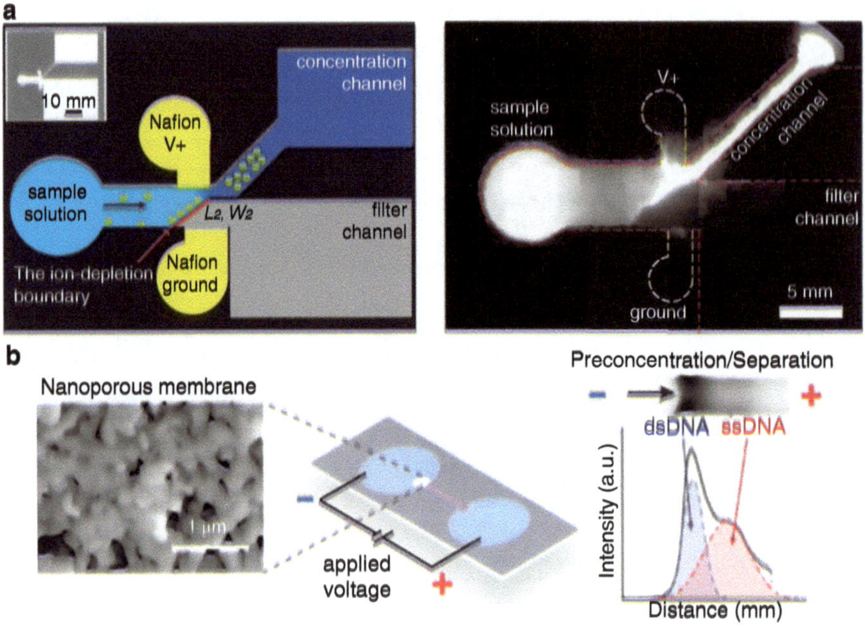

Fig. 3.9 Ion concentration polarization for preconcentration in μPADs, (**a**) separation of fluorescent particles under flow [45], (**b**) ion concentration polarization for the separation and preconcentration of ssDNA [46], reproduced with permission from references [45, 46], respectively

electroosmotic flow, as shown in Fig. 3.9b [46]. Electrokinetic stacking employs an electric field to carry out a titration between a background electrolyte and a sample solution in a μPAD. This forms a titrated/depleted zone of low conductivity in the channel, generating a high electric field at the boundary between this zone and the background electrolyte, and concentrating sample ions at this interface This has been previously employed for 150-fold stacking/preconcentration of proteins [47]. Finally, charged polyelectrolytes such as Nafion or polymer inclusion membranes have been demonstrated for the preconcentration of charged species such as Cu or synthetic dyes [48, 49]. These are simple to implement but are not as powerful at preconcentration compared to ion concentration polarization of electrokinetic stacking.

3.4 Reaction Mechanisms

While paper has many different properties that are advantageous for point-of-need detection, one of the most critical properties is its ability to store and stabilize chemical reagents. This allows for chemical and biological reactions to take place on the paper substrate, facilitating online detection of countless analytes ranging from environmental contaminants such as metals and pesticides to biological markers and infectious agents.

3.4.1 Chemical Reactions: Metal Complexation for Environmental Analysis

Electrochemistry and colorimetry are common detection approaches for the quantitation of metallic species in PADs. While metals can be directly sensed through stripping analysis, increasing sensitivity, selectivity, and lower detection limits require the use of chemical modifiers for the electrodes [50, 51]. For colorimetric metal detection, a chemical substrate is often used to complex with the metal of interest, resulting in a change of color corresponding to the concentration of the target metal (Fig. 3.10a) [5, 34, 48, 52, 53]. When developing and optimizing a PAD for metal detection based on chemical complexation, cross-reactivity with other metals is a common challenge requiring interference tests with common metals present in the matrix of interest (e.g., air, water, soil) in addition to the target metal. As well as interference test, complexation has also been used to detect organic chemicals, such as antibiotics, in water samples [54, 55].

3.4.2 Nanoparticles

Nanoparticles composed of several different materials have been used in paper-based devices, but gold and silver nanoparticles are two of the most common

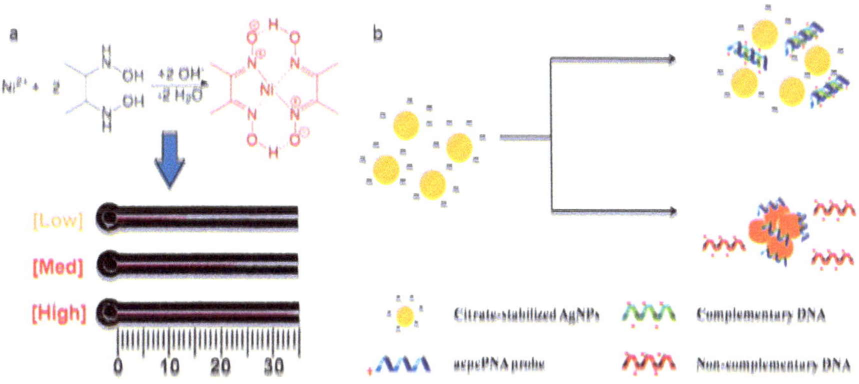

Fig. 3.10 (**a**) Distance-based detection of nickel (II) using metal complexation for a measurable color change from clear to red [52], (**b**) detecting viral DNA using peptide nucleic acids (PNA) and nanoparticles [61], reproduced with permission from references [52, 61], respectively

elements. Nanoparticles have a natural property to change color upon aggregating, making them a popular choice for simple and robust colorimetric detection on μPADs (Fig. 3.10b). Nanoparticles are associated with high extinction coefficients, facilitating a robust and sensitive method for colorimetric detection [3], although more quantifiable detection can be achieved through electrochemical detection of nanoparticles [56]. Aggregation has been used to detect metals such as copper [57] and mercury [58], along with biomolecules [59, 60], viruses [56, 61, 62], and bacteria [63]. In addition to aggregation, nanoparticles can also be used to enhance a chemical reaction [64], and silica nanoparticles have been used to enhance and unify color formation for colorimetric enzyme reactions [65].

3.4.3 Chemical Reactions for Forensic Analysis

While paper-based devices are most commonly used for environmental and biological testing, several devices have been used in point-of-need forensic applications [66]. For example, μPADs have been developed to detect compounds associated with unknown homemade explosives [67, 68]. This μPAD, which performs five separate analyses at once via a branched device, detects explosives such as TNT and RDX via detecting explosive-associated compounds including nitrate, nitrite, chlorate, perchlorate oxidizers, and ammonium. Another device has been developed for detecting unknown seized drugs for forensic analysis, including cocaine, opiates, ketamine, and phenethylamines [69]. Finally, to help combat the significant problem of adulterated whisky in Brazil, a paper-based spot test was designed to detect caramel food coloring commonly used in whisky adulteration [70].

3.4.4 Enzyme-Based Detection

Enzymes are one of the most popular detection motifs associated with paper-based devices due to their catalytic abilities to accelerate a chemical reaction and amplify a detectable product. Enzymes can be used to detect specific analytes, including one of the most popular proof-of-concept reactions used in paper-based devices: glucose [71]. This reaction is a bienzyme system that works through glucose oxidase catalyzing a reaction with glucose to result in hydrogen peroxide as a product, which is subsequently reacted with horseradish peroxidase (HRP) as seen in Fig. 3.11a. There are several substrates that can be used with HRP to result in a detectable product, either colorimetrically or electrochemically [71]. Other enzymatic reactions that have been used in μPADs include detecting other biomolecules such as uric acid, urine creatine, phenylalanine, and lactate [72–76]. In addition to enzymes' use in detecting an analyte based on color change, analytes can be detected through enzyme inhibition, resulting in a lack of color change. This has been used in detecting organophosphate pesticides that inhibit the reaction of

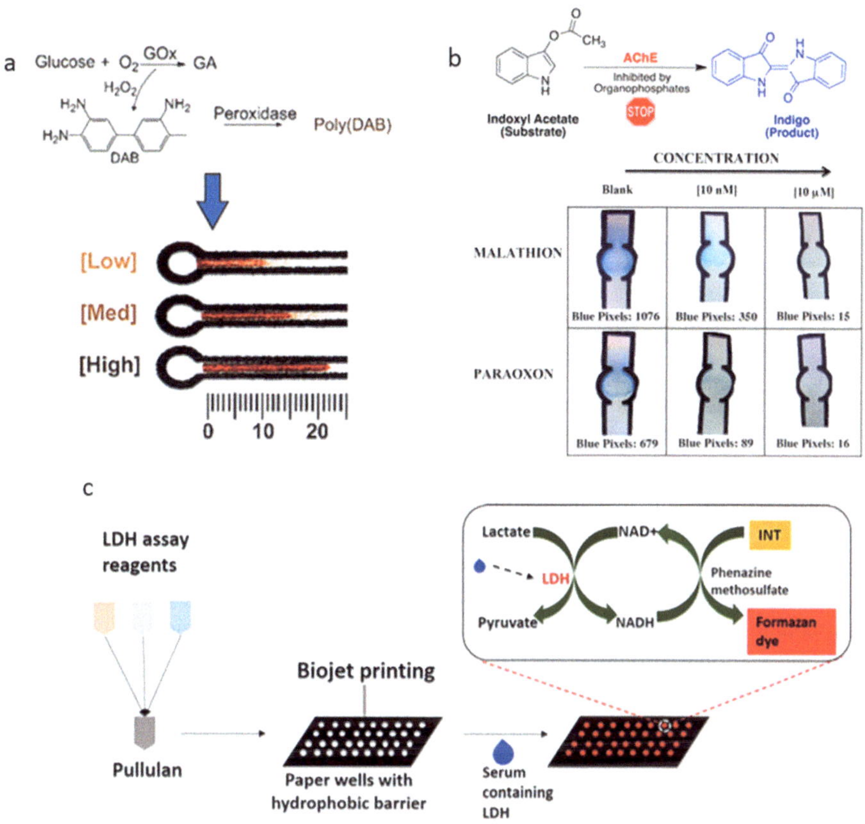

Fig. 3.11 (**a**) Using a bienzyme system of glucose oxidase (GOx) and horseradish peroxidase (HRP) to detect glucose [52], (**b**) detecting organophosphate pesticides through the inhibition of acetylcholinesterase resulting in less color change as concentration increases [77], (**c**) evaluating organ failure through the detection of naturally occurring enzyme, lactate dehydrogenase, using colorimetric substrates [86], reproduced with permission from references [52, 77, 86], respectively

acetylcholinesterase with a colorimetric substrate, resulting in less color change as the pesticide concentration increases (Fig. 3.11b) [77, 78]. For using enzymes in µPADs, several publications have discussed which paper substrates result in the most efficient reaction [79], and best chemical modifications to stabilize and store enzymes for months, increasing device robustness for field deployment [80].

In addition to using enzymes to detect analytes, substrates can be used to detect enzyme activity in a sample. Detecting the activity of specific enzymes can indicate certain pathogens or health ailments. For example, µPADs have been developed for bacteria detection based on enzymes the bacteria produce. Detecting selected enzymes that bacteria produce can distinguish between different species [81–83], pathogenicity [42], and bacteria resistant against certain antibiotics [84]. Detecting bacteria based on enzymatic activity is an attractive platform because it detects healthy and viable bacteria. Other popular methods, such as antibodies and nucleic acid amplification, detect bacteria whether the pathogens are dead or alive. Assessing human health is another application of enzyme detection, because many diseases are associated with the expression or lack of function of specific enzymes (Fig. 3.11c). Paper-based devices for health assessment have been established for detecting enzymes to evaluate liver function [85], organ failure [86], male fertility [87], and even organophosphate poisoning [88].

3.4.5 Antibody-Based Detection

Immunoassays, detecting an analyte based on antibodies, provide analytical scientists with specific and sensitive detection of biomolecules and pathogens associated with human disease. Antibodies are often very specific to the analyte with little cross-reactivity, and can be manufactured to be specific to almost any analyte. One of the most recognizable forms of paper-based devices is the pregnancy test, which is a lateral flow assay (LFA) based on antibodies specific to human chorionic gonadotropin (hCG), a protein women express when pregnant [89]. LFAs often use antibodies with covalently attached nanoparticles that aggregate together, indicating a positive sample through color change.

Another popular application of antibodies is enzyme-linked immunosorbent assays (ELISAs) where enzymes are covalently attached to antibodies instead of nanoparticles. Enzymes provide an advantage over nanoparticles because enzymes can continually amplify a substrate, providing a lower limit of detection. ELISAs are traditionally completed in 96-well polystyrene plates and a plate reader, but ELISA was introduced into paper-based devices in 2010 [90]. By moving the assay to paper, there was decreased reagent consumption, waste, and total assay time. Since its introduction to paper, ELISA has been demonstrated in the detection of biomarkers [91–93], and pathogens such as viruses [94, 95] and bacteria [96]. ELISAs are known to require a laborious procedure with several pipetting steps and reagents necessary for the final detection, making the assay less desirable for the point-of-need applications associated with μPADs. With paper's ability to store reagents, a few devices have been developed where the user adds patient sample and buffer to rehydrate reagents and interact with the sample to complete the assay. Such devices include a sliding paper-strip device [14] and a rotating device [93] where the user slides or rotates the device to deposit different reagents associated with each step of the assay. Implementing user-friendly semi-automation of the laborious procedure enables the use of ELISA as a sensitive and user-friendly method for point-of-need applications. Additionally, proper chemical modifications can facilitate storage of antibodies and enzymes on paper for at least 5 months, increasing device robustness [97]. For a detailed summary of the procedure of a sandwich-based ELISA, (see Fig. 3.12a).

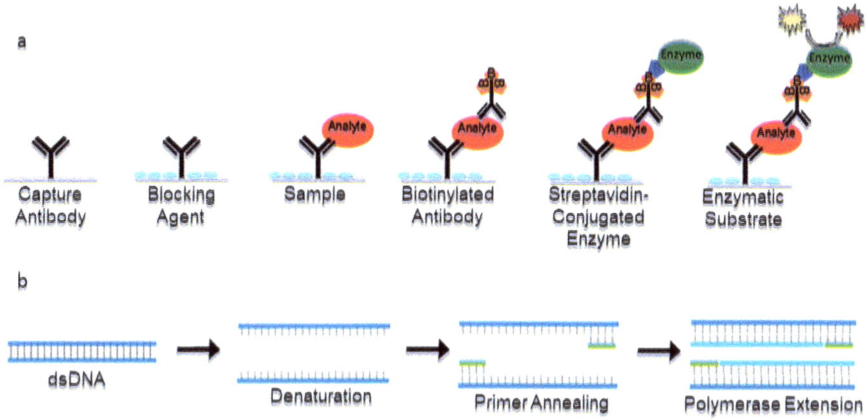

Fig. 3.12 (**a**) Nucleic acid amplification is a detection motif based on amplifying specific DNA or RNA sequences to a detectable concentration, (**b**) procedure to complete the sandwich enzyme-linked immunosorbent assay (ELISA) for specific and sensitive analyte detection

3.4.6 Nucleic Acid Amplification

One of the fastest and most sensitive diagnostic methods used in healthcare and research is the polymerase chain reaction (PCR), which is a method based on amplifying nucleic acids to a detectable concentration (Fig. 3.12a). PCR has not been directly applied to paper-based devices, likely because PCR requires a thermocycler, an expensive and complex piece of equipment that makes the assay not ideal for point-of-need settings. A thermocycler is necessary because PCR requires the solution to shift between several specific temperatures to complete the assay [98]. Instead, isothermal amplification techniques including loop-mediated isothermal amplification (LAMP) and recombinase polymerase amplification (RPA) have been applied to μPADs for sensitive detection of bacteria [99] and viruses [100–102]. Isothermal systems still require the system to be heated, only to a set temperature of 60 °C for LAMP and 40 °C for RPA. To further establish this method for field settings, an inexpensive incubator was developed using a Styrofoam cup and chemical hand warmers [103], and it was reported that polyethersulfone is the best paper substrate to use for LAMP-based μPADs [104].

3.4.7 Nucleic Acid Hybridization

Oligonucleotides, or primers, are short DNA and RNA strands that are required for nucleic acid amplification; however, they can also be used as a separate diagnostic motif to detect complementary DNA sequences. Oligonucleotides provide a specific detection motif because single-stranded DNA (ssDNA) is very specific to its own complementary sequence to form double-stranded DNA (dsDNA). These μPADs often function through a conformational change between ssDNA and dsDNA, causing a change in detectable signal (Fig. 3.12b). Peptide nucleic acids (PNAs) have also been used in oligonucleotide-based diagnostic assays [61, 105, 106]. PNA is like DNA but with a peptide-based backbone, which eliminates a negatively charged molecule and the concern of naturally occurring enzymes targeting the oligonucleotides, which are considered foreign DNA, in a biological matrix. Complementary nucleic acid sequences have been used in μPADs for the detection of nucleic acid sequences associated with cancer [106] and neurological diseases [107], and in the diagnosis of viral infections [61, 105].

3.5 Detection

Since the ultimate goal of developing PADs from the inception was to apply for POC analysis, detection schemes coupled with PADs have to be simple, rapid, sensitive, specific, cost effective, and reliable. Undoubtedly, colorimetry is the most common detection method due to its simplicity and potential for instrument-free quantification. Color change caused by a reaction occurring on PADs can be easily observed by naked eyes and corresponds to the amount of the target analyte. Alternatively, electrochemical detection is also used as a detection approach for PADs, namely electrochemical paper-based analytical devices (ePADs), to improve sensitivity, selectivity, and dynamic range of quantitative results. In addition, other detection methods, such as chemiluminescence (CL), electrochemiluminescence (ECL), fluorescence (FL), and Raman spectroscopy have also been applied for detection of the assays on PADs.

3.5.1 Colorimetry

Colorimetric detection has proved as the most common detection method for PADs due to its simplicity, reliability, low cost, and high possibility for instrument-free detection. This detection method is based on visually observing the change in color intensity or brightness caused by an enzymatic or chemical reaction and comparing the change with the standards with known color intensity. Previously, colorimetric detection as a "yes/no" answer or semiquantitative readout can be simply achieved by naked eyes, which is instrumental-free detection, and is sufficient for analysis. However, differences in perception of people and external lighting conditions can cause variation in color intensity. To obtain more reliable quantitative results, an imaging tool (e.g., a digital camera, a phone camera, or a scanner) is used to capture images of the reaction and the images are further analyzed by imaging software (e.g., Illustrator, ImageJ) to accurately measure the intensity which corresponds to the amount of analyte. Accurately quantifying colorimetric assays on PADs can be achieved by measuring the amount of light reflected from the assays on paper using an appropriate imaging tool, e.g., a digital camera, a mobile phone camera/smartphone, or a scanner.

Scanner: An office scanner is a common device for measuring color intensity of the assays on PADs. A scanner provides high resolution, is user friendly, and is not affected by the external lighting conditions [2]. In addition, all imaging parameters for scanning can be set accurately. However, measuring color intensity using a scanner requires a dried reaction zone. Accordingly, the colorimetric assay on PADs is normally air-dried before being scanned by a scanner, which could take longer than an hour, depending on the assay volume. In addition, differences in the color of the assay between a dry paper and a wetted paper could cause variation of results. Moreover, a problem associated with color fade while the reaction is drying out could result in low sensitivity and a narrow dynamic range. Furthermore, color analysis of toxic substances or biomarkers of infectious diseases using a scanner without proper cleaning and disinfection could cause contamination and spread out diseases.

Digital camera: A digital camera is an alternative and widely used to capture images of colorimetric assays on PADs. Digital cameras with CCD sensors provide high-quality image data. Using a digital camera, images of the assay can be captured at any reaction time from the beginning of the reaction until it dries out. Normally, the reaction time is fixed for taking images to ensure that the color is measured at the same reaction time to solve the problem associated with color fading with time. By capturing images, there is no direct contact between a camera and the reaction, preventing cross-contamination and infection. In addition, all imaging parameters of the camera can be set and controlled precisely and accurately. Images with high quality can be obtained with digital cameras, which is suitable for applications that require high-resolution images. Unlike scanners, capturing images using digital cameras requires external illumination. Accordingly, light setting for photograph is important because it can affect the intensity of digital images. In addition, the focus of the image relies on the photographers; however, autofocus can be used but it is less reliable when compared to scanners. With compact and portable devices, digital cameras have gained popularity for imaging the colorimetric assays on PADs.

Mobile phone camera: Nowadays, mobile phones are undoubtedly important to our daily life. The advanced technology brings the price of mobile phones to be much cheaper than before, thus making it affordable for everyone. In addition, other features/functions, such as cameras, are significantly improved. Therefore, with their lightweight, portability, low cost, and user-friendly devices, mobile phones promise potentially imaging tools for colorimetric detection of PADs. Although image quality generally obtained from mobile phone cameras with CMOS sensors is not as high as that of digital cameras, it is sufficient for color analysis of some applications with less quality demands. Similar to digital cameras, lighting conditions affect the intensity of images. In addition, some mobile phones have a problem with a long focal distance, which makes the phone unable to capture close-up images to have high resolution of small test areas. This can be alleviated by using additional lens to help focusing on close objects. Moreover, angles of imaging and shadow also cause variations in the quality of images taken by camera phones. However, the key advantage in using mobile phone cameras as imaging tools is that images taken on-site or remoted areas can be transferred electronically afterwards to experts or physical doctors off-site to interpret results for diagnostics and medical treatments. Accordingly, mobile phones offer a direct route to be useful tools in telemedicine.

Image analysis: Images of the assays captured by imaging tools are further processed using image manipulation software, such as Adobe Photoshop® or ImageJ, which is commercially available to obtain color information for quantitative analysis. Region of interest of the colorimetric assays on PADs is defined for color quantification. Color changes or the change in color intensity is quantified using a color system, which is also known as a color space. Normally, color spaces specify color based on the three primary colors, which is similar to the human visual system. Therefore, most of the color spaces are represented as three components [108], such as RGB (red, green, and blue), HSV (hue, saturation, and value), and CIE (the International Commission on Illumination) 1931. Among these color spaces, RGB is the most common color space used in imaging devices and color detectors and widely used in digital processing of images to extract color information because it works in a similar way as the human visual system [108]. In addition, the development of other color spaces is based on the RGB system. Most of the colorimetric PADs use the RGB system for color quantification due to its simplicity and direct measurements. However, RGB is a device-dependent color system [108]. Therefore, different devices could provide color information based on RGB system differently. Accordingly, when working at high-resolution applications, RGB could be problematic and other color spaces might be considered instead. For example, HSV [109] and CIE 1931 [110] are also applied for color analysis in colorimetric PADs to minimize the effect of light variations in the illumination, resulting in more accurate and reproducible measurements. Apart from three component representations, other color spaces, such as CYMK (representing cyan, yellow, magenta, and black) which is normally used for color printing, have been used in colorimetric analysis for PADs to obtain a larger dynamic range and improve precision in the color analysis for some applications [111, 112]. Accordingly, selection of color spaces for color quantification depends on many factors, such as applications, illumination, color change, and instrument used [108]. However, RGB is normally chosen first because it is simple and more familiar with users. If the RGB system is not satisfied, other color systems are considered to determine the best fit model.

3.5.2 Electrochemical Detection

Analysis by capturing images and measuring color intensity relies on selected background and environmental illumination, imaging devices, and photographers, resulting in irreproducible and unreliable results. In addition, a narrow dynamic range and low sensitivity are typically observed from colorimetric detection [51]. To overcome these limitations associated with colorimetric detection, an electrochemical sensor is an alternative to be used as a detection approach for PADs. Electrochemical paper-based analytical devices (ePADs) have shown the improvements in terms of sensitivity and selectivity with slightly more complicated systems compared to colorimetric PAD formats. However, a portable potentiostat renders ePADs applicable for on-site monitoring. In addition, electrodes can be minimized and easily fabricated on paper. Accordingly, electrochemical sensing has shown its ability to be coupled as a good match for PADs.

Electrode fabrication methods: Based on electrode fabrication of microfluidic systems, electrodes for PADs have followed the footprints of those used in traditional microfluidic formats. There are a variety of electrode fabrication methods on paper (Fig. 3.13), such as screen/stencil printing [113–115], ink-jet printing [116, 117], drawing [118–121], microwire [122, 123], and sputtering [124–127]. Among the emerging techniques, screen-printed electrodes (SPEs), first reported as electrodes for ePADs [113], have gained popularity due to their ease of fabrication on paper, high reproducibility in terms of batch-to-batch fabrication, low cost, and suitability for mass production [114]. In addition, SPEs can be easily modified with a variety of recognition elements, such as enzymes [113, 115], antibodies [56, 128–130], aptamers [131], and molecular imprinting polymers [132, 133]. To enhance the electrochemical performance of SPEs, modifiers, such as conducting polymers [128, 132], carbon nanomaterials (nanotubes [134, 135], graphene [135, 136], and carbon back [135, 137]), mediators (tetrathiafulvalene [134] and Prussian blue [138]), and gold nanoparticles [139, 140], have been used for electrode modification. Nowadays, both modified and bare SPEs have been commercially available in a variety of shapes and designs [141].

Different electrochemical approaches, such as voltammetry, amperometry, and potentiometry, have been applied for ePADs, depending on applications and sensitivity. Voltammetry, especially anodic striping voltammetry (ASV), has been demonstrated as a highly sensitive method for trace analysis of heavy metals due to its high sensitivity, selectivity, and reproducibility [51]. ASV involves a preconcentration step in which a metal ion of interest is electrodeposited on the working electrode surface and a stripping step in which the deposited metal is selectively swept from the electrode by applying an anodic potential sweep for the oxidation process. The current, corresponding to the analyte concentration, is measured during the stripping process. ePADs coupled with ASV have been reported for highly sensitive and selective determination of a variety of metals, such as Pb^{2+}, Cd^{2+}, Zn^{2+}, Hg^{2+}, Cr^{3+}, Ag^{+}, and Au^{3+} [142, 143].

Amperometric detection is the measurement of current from the oxidation/reduction of an electroactive species. The current obtained from the reaction is proportional to the analyte concentration. Compared with other amperometric techniques, chronoamperometry, measuring current as a function of time at a constant applied voltage, offers a better signal-to-noise ratio because the current is integrated over a period of time [115]. Chronoamperometry measures Faradaic

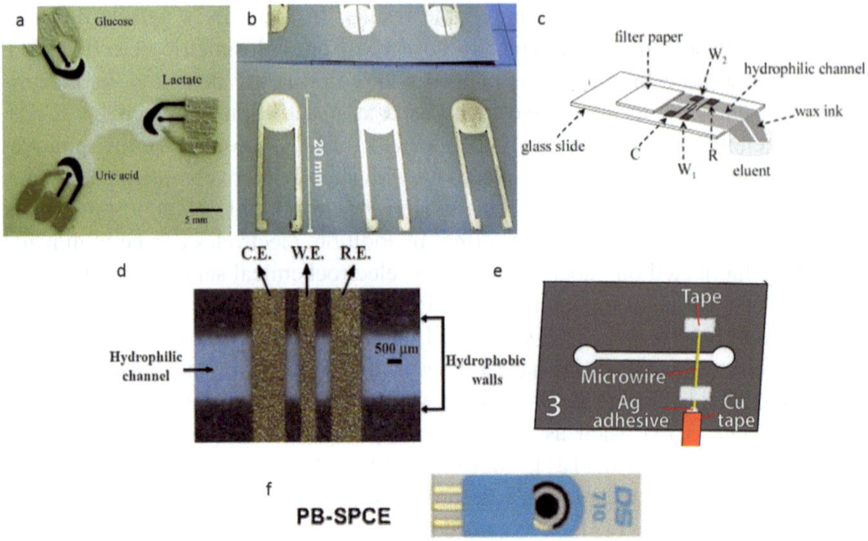

Fig. 3.13 Electrodes of ePADs fabricated using different methods; (**a**) screen printing [113], (**b**) ink-jet printing [116], (**c**) pencil drawing [118], (**d**) sputtering [124], (**e**) a microwire [122], and (**f**) a commercial screen-printed carbon electrode (SPCE) modified with Prussian blue (PB) [161], reproduced with permission from references [113, 122, 161], respectively

current (i in unit A) which is proportional to the concentration of the analyte according to the Cottrell equation ($i = nFAD^{1/2}\pi^{-1/2}\, t^{-1/2}C$), where n is the number of electron, t is the time (s), F is Faraday's constant (96,485 C mol^{-1}), A is the area of the electrode (cm^2), D is the diffusion coefficient of the analyte ($cm^2\ s^{-1}$), and C is the initial concentration of the reactant (mol cm^{-3}) [115]. Amperometric measurements are normally used with ePADs because most of the assays are based on enzymes which catalyze the oxidation or reduction of the analyte of interest into an electroactive species [115, 144]. The most common example of amperometric measurements on ePADs is determination of glucose in biological samples, such as urine, blood, and sweat.

Potentiometric detection, measuring the potential difference between an indicator electrode and a reference electrode, is applied in ePADs for determination of ions through the use of ion-selective electrodes (ISEs) [145, 146]. The use of ISEs allows for selective analysis at low cost. In addition, the paper itself also acts as a filter which could prevent large particles/molecules, especially when performing measurements in biological samples, from reaching the electrode surface, minimizing electrode fouling. Due to the difficulty of fabricating ISEs on paper and electrode fouling, ePADs with potentiometric measurements are normally applied for determination of inorganic ions, such as Na^+, K^+, Ca^{2+}, Cd^{2+}, Pb^{2+}, Cl^-, and pH [145, 146].

3.5.3 Luminescence

Luminescence detection approaches, such as chemiluminescence (CL), electrochemiluminescence (ECL), and fluorescence (FL), have attracted considerable attention to be coupled with PADs. CL and ECL, measuring the emission of light generated from a chemical reaction and an electrochemical reaction, respectively, are simple, inexpensive, and portable and require less instrumentation. The advantages of ECL over CL and FL are that electrode potential is controlled more easily than reagent addition at a specific time and selectivity of the method can be achieved by controlling the electrode potential [3]. Unlike CL and ECL, FL requires an excitation source to obtain luminescence, rendering the technique more complicated in terms of extra instruments. In addition, paper-whitening additives could increase the background fluorescence [3]. However, CL, ECL, and FL offer more sensitive detection methods for PADs when compared to the colorimetric assays. Therefore, improved detection limits are normally achieved when performing assays based on CL [147], ECL [148], and FL [149].

3.5.4 Other Detection Methods

Raman spectroscopy is also applied as a detection approach for PADs. Surface-enhanced Raman spectroscopy (SERS) has emerged as a powerful technique providing structural information of molecules with high detection sensitivity. The miniaturization of Raman spectrometers enables the integration of SERS and PADs for on-site applications. There is currently growing interest in fabrication of low-cost, facile, easy, and disposable paper-based SERS sensors with excellent SERS performance [150]. Moreover, mass spectrometry (MS) as a powerful detection method has been incorporated with PADs for the analysis of complex samples to provide chemical structures and molecular weights of analytes [151].

3.6 Result Output

For colorimetric detection, result readout can be categorized into three different approaches: (i) qualitative output, (ii) semiquantitative output, and (iii) quantitative output. However, quantitative results are normally obtained from ePADs with electrochemical signal.

3.6.1 Colorimetric Detection

Qualitative outputs: The simplest output obtained from colorimetric PADs is whether the analyte is present in the sample, which is a yes/no answer. This approach requires no instrument because a change of color due to the presence of the analyte of interest is observed by naked eyes from the reaction zone. However, a control is needed to prevent a fault positive and confirm that the reaction is working properly. Accordingly, this type of result display is normally found with dipsticks and lateral flow assays which require rapid diagnostics and allow for people with little to no training to perform the tests themselves [152]. The most well-known yes/no assay performed on a lateral flow format is a pregnancy test. However, a yes/no assay can be found in other applications, as shown in Fig. 3.14 which is an example of a test strip for detection of pesticides.

Quantitative results are important when a “yes” or “no” readout is not sufficient. Therefore, most of the assays on paper are aimed to quantify the amount of the

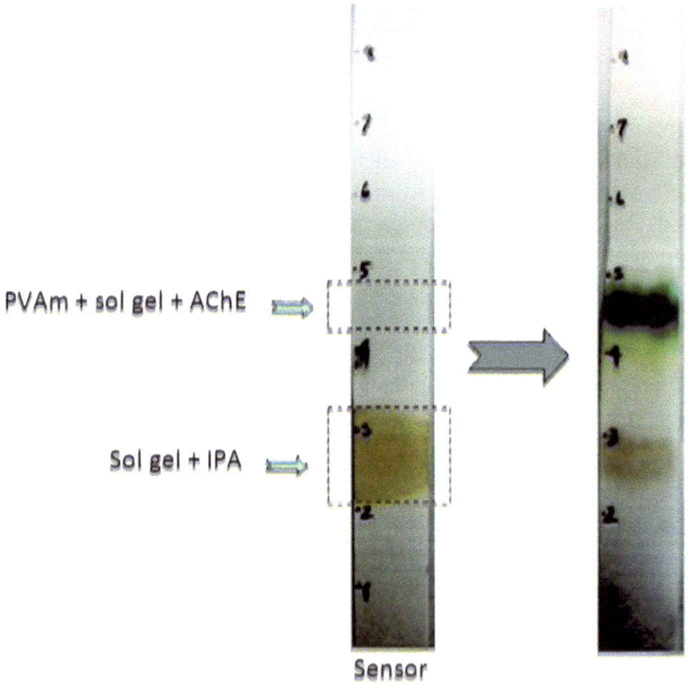

Fig. 3.14 Lateral flow bioactive paper sensors for detection of pesticides. The detection mechanism is based on the reaction between acetylcholinesterase (AChE) and indophenyl acetate (IPA), a chromogenic substrate, yielding a blue color. Upon the presence of pesticides, which are AChE inhibitors, the disappearance of the blue color is observed. Polyvinyl amine (PVAm), a cationic polymer, is used to trap the blue product to be a confined zone [78], adapted with permission from reference [78]

analyte of interest. However, with the concept of PADs originally developed to be an instrumental-free testing device, little to no instrument is preferred for quantification of analytes, leading to semiquantitative colorimetric analysis. To be able to perform semiquantitative analysis with no external instrument, a color-chart scale/bar is normally provided with the paper devices as a color read guide for the comparison between the color obtained from the reaction on the PADs and the color-chart scale to estimate the amount of the analyte. Therefore, only a concentration range or an approximate concentration of the analyte is obtained, which is not highly accurate. However, in some circumstances, semiquantitative readouts with no instrument may be sufficient to differentiate the amount of analyte to low, moderate, and high levels, which can be used to classify the patient's conditions into three different groups when performing screening test for medical diagnostics in remote areas with low resources, as shown in Fig. 3.15.

Normally, performing assays on PADs requires precisely quantitative results; therefore, a concentration range or an approximate concentration of the target analyte obtained from semiquantitative readouts is not sufficient for many applications. An external instrument is normally required to obtain quantitative results from PADs; however, quantitative analysis on PADs is also possible without extra equipment. To perform quantitative analysis on PADs, an external instrument, such as scanners, digital cameras, or smartphone cameras, is normally involved

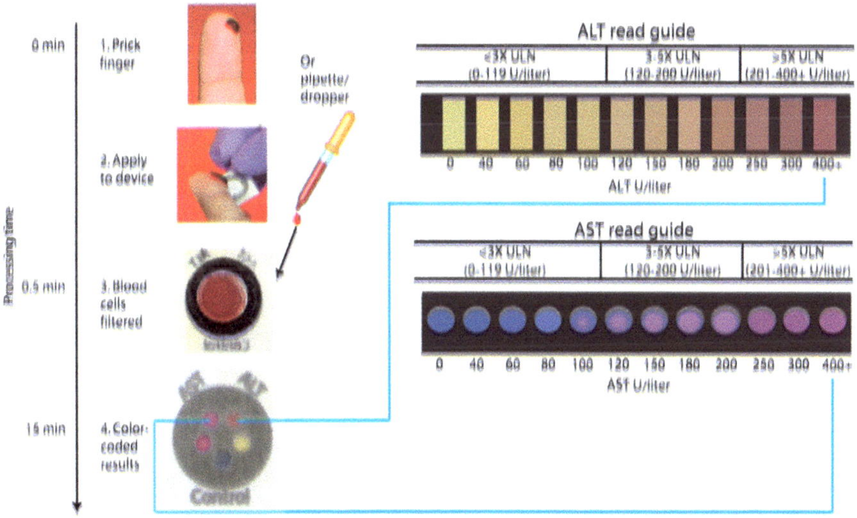

Fig. 3.15 A multiplexed paper-based device for monitoring drug-induced liver injury through semiquantitative measurements of serum transaminases (aspartate aminotransferase (AST) and alanine aminotransferase (ALT)). Blood/serum samples are applied to the test zones of a paper device for individual testing of AST and ALT. The colors developed from the assays are compared with the provided color read guide to classify the patient's condition into three different regions (<3× upper limit of normal (ULN), 3–5× ULN, and >5× ULN) with >90% accuracy [85], reproduced with permission from reference [85]

for imaging the colorimetric assays before using imaging software to measure color intensity associated with the amount of the target analyte. Generally, a calibration curve is constructed by plotting color intensity as a function of analyte concentration. Subsequently, the concentration of the analyte is extrapolated from the calibration curve using its color intensity. This method provides more accurate readouts compared to semiquantitative methods.

To make it simple and practical for working in remote areas with low resources, an instrument-free approach is also developed for quantitative analysis using PADs. This can be achieved when using PADs in a distance-based platform. Distance-based detection, an alternative for quantitative readout, can be employed for quantitative analysis on PADs without the assistance of an external instrument [52]. This type of paper devices consists of a circular testing zone connected with a straight channel with a printed scale bar for measuring a distance (Fig. 3.16a). Instead of measuring color intensity developed from the assay, color length travelling along the straight channel is read out from the printed scale, which corresponds to the amount of the target analyte. This detection approach enables instrument-free quantitative measurements on PADs and addresses the classical problems associated with background selection and environmental illumination of colorimetric PADs. Both semiquantitative and quantitative measurements can be obtained from distanced-

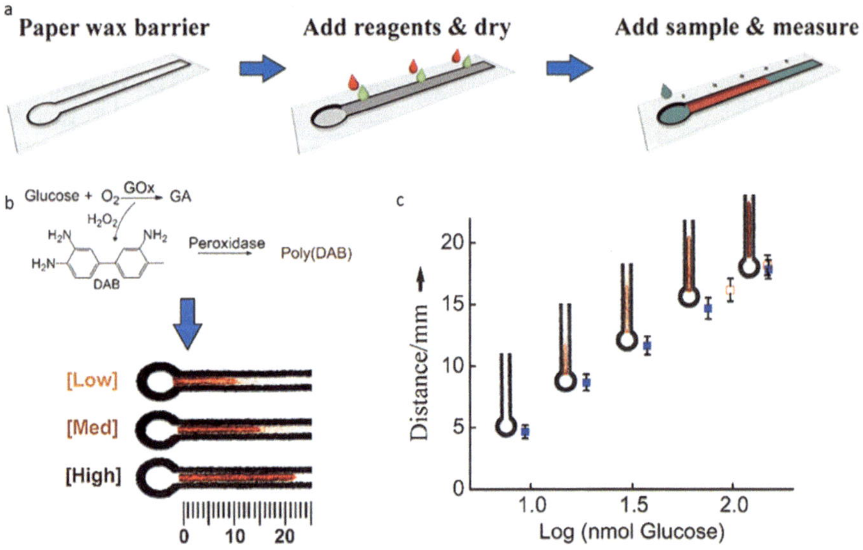

Fig. 3.16 (**a**) An operational concept of distance-based detection for quantitative measurements. (**b**) A distance-based detection approach for semiquantitative and quantitative measurements of glucose using PADs. Brown color developed along the channel is the product from the reaction between H_2O_2 (generated from the oxidation of glucose by glucose oxidase) and 3,3′-diaminobenzidine (*DAB*) with the assistance of peroxidase enzyme. (**c**) A calibration curve plotted between color distance and glucose concentration is used for quantitative measurements [52], reproduced with permission from reference [52]

based detection. A calibration curve constructed from the color distance as a function of analyte concentration is normally employed to obtain more accurate results for quantitative analysis (Figure 3.16c).

3.6.2 Electrochemical Detection

ePADs are intrinsically instrument-based detection methods. Electrochemical signal obtained from the assay is processed and displayed in a digital format. Accordingly, quantitative results are normally received from ePADs. Similar to colorimetric detection, a calibration curve, plotted between electrochemical readout and concentration, is normally constructed to obtain more accurate results.

3.7 Digital Readers

Normally, a process to obtain quantitative results involves detection of assays, constructing a calibration curve, and extrapolating concentration of the target analyte from the calibration curve. All these processes are performed manually, and thus time consuming. To expedite all the processes, extra devices as assay readers are developed to automatically detect, process, and transfer all assay information into a digital format and the concentration of analyte can be displayed on a screen straightaway after the detection process. Therefore, assay readers consist of detection tools and software for detecting and digitizing the assays.

3.7.1 Handheld Readers

Handheld readers are developed to be incorporated with PADs to digitize the signal from the assays (Fig. 3.17). For colorimetric assays, handheld readers are developed based on optical measurements of either the transmission of light through paper [153] or the reflectance of light [154]. After the assays, a paper device is directly inserted into a handheld reader for signal processing and results are displayed on the screen. In addition, handheld readers can transfer digital information over a smartphone which can further transmit the information to experts. However, the handheld readers are customized to be suited with a particular PAD. For electrochemical detection, ePADs can be easily integrated with handheld readers. Taking advantages from the existing technology for POC diagnostics, the commercial POC devices can be adapted to be used as handheld readers for ePADs. For example, a commercial glucometer can be coupled with ePADs for quantitative analysis of compounds (e.g., glucose, cholesterol, lactate, and alcohol) relevant to human health [155]. However, handheld readers encounter difficulties associated

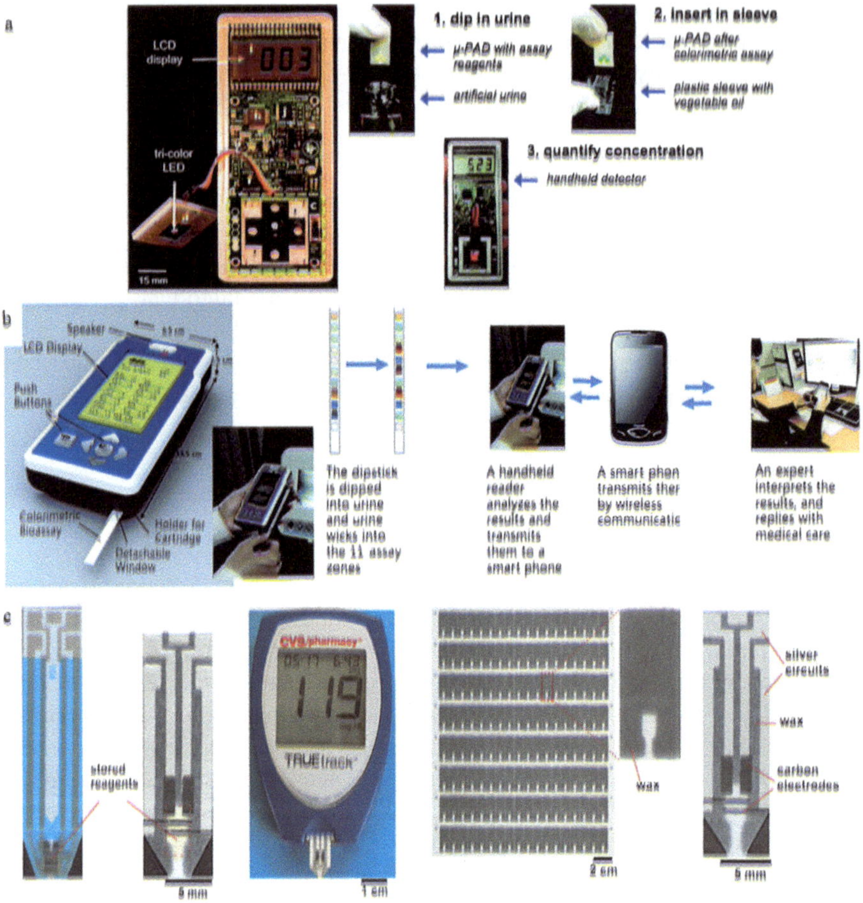

Fig. 3.17 Handheld readers for PADs. For colorimetric assays, handheld optical colorimeters are developed for quantifying colorimetric assays by means of (**a**) transmittance-based measurements [153] or (**b**) reflectance-based measurements [154], (**c**) a commercial glucometer employed as a handheld reader for ePADs to quantitatively measure glucose, cholesterol, lactate, and alcohol in blood and urine [155], reproduced with permission from references [153–155], respectively

with transferring information because they are not wireless communication devices. Alternatively, an external reader is developed into a compact platform to attach on a mobile phone to digitize information obtained from images taken by the phone camera (Fig. 3.18). Images captured by a mobile phone camera are further processed through a customized application running on a mobile phone to generate all quantitative results and display digitally afterwards.

Similar to colorimetric assays, smartphones are also developed for electrochemical analysis. Based on potentiometric measurements, the potential arising from the reaction can be measured by a potentiometric reader connected to a smartphone via a USB port (Fig. 3.19) [156]. A customized application of the smartphone has

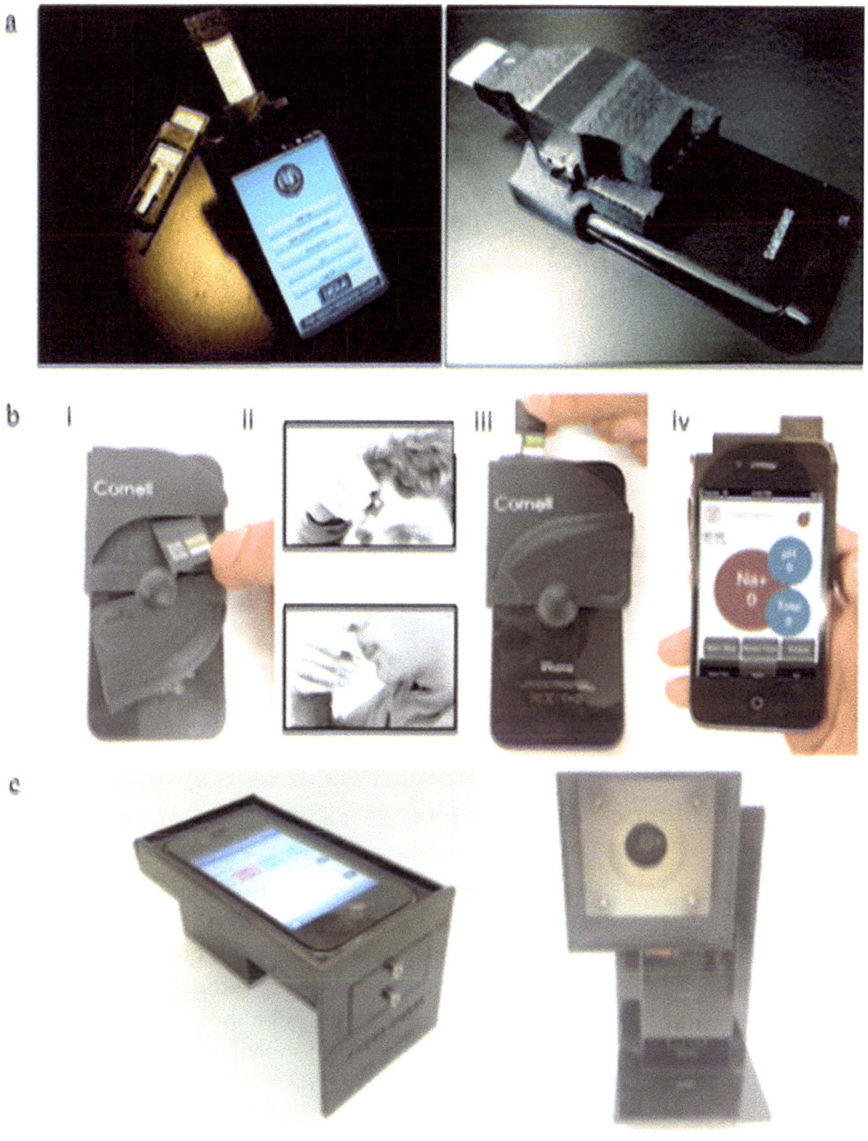

Fig. 3.18 External readers coupled with smartphones for colorimetric PADs; (**a**) an optical reader installed on an Android phone (Samsung Galaxy S II) for rapid diagnostics of malaria, tuberculosis, and HIV [162]; (**b**) an integrated smartphone accessory for monitoring changes in sweat and salivary pH [163]: (i) a smartphone case having a storage compartment for test strips, (ii) sampling methods for sweat and saliva samples, (iii) insertion of a test strip for data acquisition and reading, (iv) results are analyzed using a specially developed iPhone app and displayed on the screen; (**c**) the combination between colorimetric PADs and a smartphone for determination of explosives [164], reproduced with permission from references [162–164], respectively

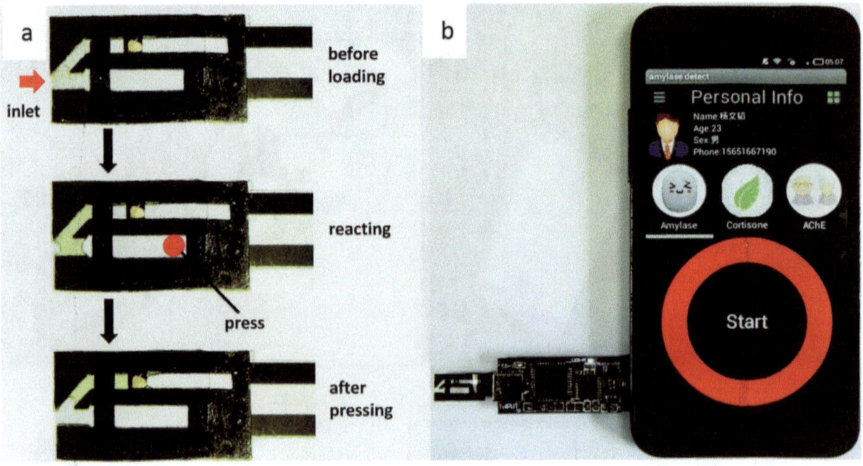

Fig. 3.19 Potentiometric measurements of salivary α-amylase using a smartphone [156], (**a**) a testing device consists of electrodes screen-printed onto a transparency film, (**b**) a potentiometric reader with a testing device is connected to a smartphone via a USB port to convert electrochemical signal into analyte concentration, reproduced with permission from reference [156]

been developed to convert electrochemical signal from the potentiometric reader into analyte concentration based on a calibration curve.

3.7.2 Smartphone-Based Readers

The key advantage of smartphone-based readers is that after the assays, all information from PADs can be directly transferred to experts or medical doctors for data interpretation or medical treatments. In addition, mobile phones allow for data sharing, providing real-time results and storing information in a global database in case of monitoring evolution of diseases. Moreover, some applications of mobile phones have functions to extract color information directly from JPEG images. However, to obtain more accurate results, a customized mobile phone application is properly developed to perform image acquisition and processing and handle with data storage and manipulation. This makes smartphones having dual functions a capture device to detect color from the colorimetric paper-based assays and a reader. Accordingly, this type of smartphone consists of a built-in camera and an internal microprocessor for imaging analysis and a data-processing unit, avoiding the need of external computers and instruments and providing fast and accurate results (Fig. 3.20). Although smartphones have shown their ability for mobile health applications and the combination of smartphones and PADs could offer a potential platform for disease screening in developing countries and remoted areas with limited resources, they are still limited to some models of smartphones

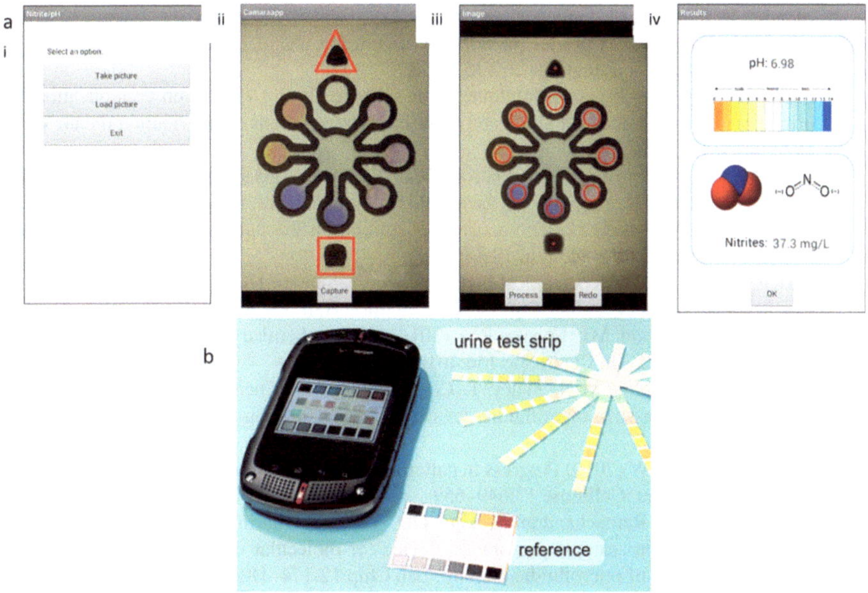

Fig. 3.20 Smartphone-based readers for colorimetric PADs, (**a**) a mobile phone with an android application for colorimetric measurements of nitrite concentration and pH determination [109]: (i) main menu, (ii) acquisition process, (iii) marks and selected areas for detection, and (iv) result displays on the screen after imaging analysis and data processing; (**b**) a smartphone for pH measurements using colorimetric urine test strips [110], reproduced with permission from references [109, 110], respectively

and software or custom-developed applications are specially needed to perform the assays, implying that they are yet to gain popularity.

References

1. Martinez AW, Phillips ST, Butte MJ, Whitesides GM (2007) Patterned paper as a platform for inexpensive, low-volume, portable bioassays. Angew Chem Int Ed 46:1318–1320
2. Nery EW, Kubota LT (2013) Sensing approaches on paper-based devices: a review. Anal Bioanal Chem 405:7573–7595
3. Cate DM, Adkins JA, Mettakoonpitak J, Henry CS (2015) Recent developments in paper-based microfluidic devices. Anal Chem 87:19–41
4. Channon RB, Nguyen M, Scorzelli A, Henry E, Volckens J, Dandy D, Henry C (2018) Rapid flow in multilayer microfluidic paper-based analytical devices. Lab Chip 18:793–802
5. Cate DM, Noblitt SD, Volckens J, Henry CS (2015) Multiplexed paper analytical device for quantification of metals using distance-based detection. Lab Chip 15:2808–2818
6. Singer JM, Plotz CM (1956) The latex fixation test. Am J Med 21:888–892
7. Washburn EW (1921) The dynamics of capillary flow. Phys Rev 17:273–283

8. Mendez S, Fenton EM, Gallegos GR, Petsev DN, Sibbett SS, Stone HA, Zhang Y, López GP (2010) Imbibition in porous membranes of complex shape: quasi-stationary flow in thin rectangular segments. Langmuir 26:1380–1385
9. Darcy H (1856) Les fontainespubliques de la ville de Dijon : exposition et application des principes à suivre et des formules à employer dans les questions de distribution d'eau. Victor Dalmont, Paris, p 647
10. Camplisson CK, Schilling KM, Pedrotti WL, Stone HA, Martinez AW (2015) Two-ply channels for faster wicking in paper-based microfluidic devices. Lab Chip 15:4461–4466
11. Hong S, Kim W (2015) Dynamics of water imbibition through paper channels with wax boundaries. Microfluid Nanofluid 19:845–853
12. Larich VG (1962) Physicochemical hydrodynamics, Prentice-Hall international series in the physical and chemical engineering sciences. Prentice-Hall, Englewood Cliffs, NJ, p 700
13. Renault C, Anderson MJ, Crooks RM (2014) Electrochemistry in hollow-channel paper analytical devices. J Am Chem Soc 136:4616–4623
14. Verma MS, Tsaloglou M-N, Sisley T, Christodouleas D, Chen A, Milette J, Whitesides GM (2018) Sliding-strip microfluidic device enables ELISA on paper. Biosens Bioelectron 99:77–84
15. Li X, Tian J, Shen W (2010) Progress in patterned paper sizing for fabrication of paper-based microfluidic sensors. Cellulose 17:649–659
16. Govindarajan AV, Ramachandran S, Vigil GD, Yager P, Bohringer KF (2012) A low cost point-of-care viscous sample preparation device for molecular diagnosis in the developing world; an example of microfluidic origami. Lab Chip 12:174–181
17. Cai L, Zhong M, Li H, Xu C, Yuan B (2015) Defining microchannels and valves on a hydrophobic paper by low-cost inkjet printing of aqueous or weak organic solutions. Biomicrofluidics 9:046503
18. Koo CKW, He F, Nugen SR (2013) An inkjet-printed electrowetting valve for paper-fluidic sensors. Analyst 138:4998–5004
19. Li X, Zwanenburg P, Liu X (2013) Magnetic timing valves for fluid control in paper-based microfluidics. Lab Chip 13:2609–2614
20. Giokas DL, Tsogas GZ, Vlessidis AG (2014) Programming fluid transport in paper-based microfluidic devices using razor-crafted open channels. Anal Chem 86:6202–6207
21. Toley BJ, McKenzie B, Liang T, Buser JR, Yager P, Fu E (2013) Tunable-delay shunts for paper microfluidic devices. Anal Chem 85:11545–11552
22. Lutz B, Liang T, Fu E, Ramachandran S, Kauffman P, Yager P (2013) Dissolvable fluidic time delays for programming multi-step assays in instrument-free paper diagnostics. Lab Chip 13:2840–2847
23. Li CG, Joung H-A, Noh H, Song M-B, Kim M-G, Jung H (2015) One-touch-activated blood multidiagnostic system using a minimally invasive hollow microneedle integrated with a paper-based sensor. Lab Chip 15:3286–3292
24. Liu H, Li X, Crooks RM (2013) Paper-based SlipPAD for high-throughput chemical sensing. Anal Chem 85:4263–4267
25. Chen D, Mauk M, Qiu X, Liu C, Kim J, Ramprasad S, Ongagna S, Abrams WR, Malamud D, Corstjens PLAM, Bau HH (2010) An integrated, self-contained microfluidic cassette for isolation, amplification, and detection of nucleic acids. Biomed Microdevices 12:705–719
26. Martinez AW, Phillips ST, Nie Z, Cheng C-M, Carrilho E, Wiley BJ, Whitesides GM (2010) Programmable diagnostic devices made from paper and tape. Lab Chip 10:2499–2504
27. Niedl RR, Beta C (2015) Hydrogel-driven paper-based microfluidics. Lab Chip 15:2452–2459
28. Houghtaling J, Liang T, Thiessen G, Fu E (2013) Dissolvable bridges for manipulating fluid volumes in paper networks. Anal Chem 85:11201–11204
29. Apilux A, Ukita Y, Chikae M, Chailapakul O, Takamura Y (2013) Development of automated paper-based devices for sequential multistep sandwich enzyme-linked immunosorbent assays using inkjet printing. Lab Chip 13:126–135

30. Nguyen MP, Meredith NA, Kelly SP, Henry CS (2018) Design considerations for reducing sample loss in microfluidic paper-based analytical devices. Analytica Chimica Acta 1017: 20–25
31. Dharmaraja S, Lafleur L, Byrnes S, Kauffman P, Buser J, Toley B, Fu E, Yager P, Lutz B (2013) Programming paper networks for point of care diagnostics. SPIE MOEMS-MEMS 8615:11
32. Fu E, Liang T, Spicar-Mihalic P, Houghtaling J, Ramachandran S, Yager P (2012) Two-dimensional paper network format that enables simple multistep assays for use in low-resource settings in the context of malaria antigen detection. Anal Chem 84:4574–4579
33. Fu E, Downs C (2017) Progress in the development and integration of fluid flow control tools in paper microfluidics. Lab Chip 17:614–628
34. Rattanarat P, Dungchai W, Cate D, Volckens J, Chailapakul O, Henry CS (2014) Multilayer paper-based device for colorimetric and electrochemical quantification of metals. Anal Chem 86:3555–3562
35. Meredith NA, Quinn C, Cate DM, Reilly TH 3rd, Volckens J, Henry CS (2016) Paper-based analytical devices for environmental analysis. Analyst 141:1874–1887
36. Meredith NA, Volckens J, Henry CS (2017) Paper-based microfluidics for experimental design: screening masking agents for simultaneous determination of Mn(ii) and Co(ii). Anal Methods 9:534–540
37. Bhamla MS, Benson B, Chai C, Katsikis G, Johri A, Prakash M (2017) Hand-powered ultralow-cost paper centrifuge. Nat Biomed Eng 1:0009
38. Yang X, Forouzan O, Brown TP, Shevkoplyas SS (2012) Integrated separation of blood plasma from whole blood for microfluidic paper-based analytical devices. Lab Chip 12:274–280
39. Samad Khan M, Thouas G, Shen W, Whyte G, Garnier G (2010) Paper diagnostic for instantaneous blood typing. Anal Chem 82:4158–4164
40. Songjaroen T, Dungchai W, Chailapakul O, Henry CS, Laiwattanapaisal W (2012) Blood separation on microfluidic paper-based analytical devices. Lab Chip 12:3392–3398
41. He J, Huang M, Wang D, Zhang Z, Li G (2014) Magnetic separation techniques in sample preparation for biological analysis: a review. J Pharm Biomed Anal 101:84–101
42. Zakir Hossain SM, Ozimok C, Sicard C, Aguirre SD, Monsur Ali M, Li Y, Brennan JD (2012) Multiplexed paper test strip for quantitative bacterial detection. Anal Bioanal Chem 403:1567–1576
43. Srisa-Art M, Boehle KE, Geiss BJ, Henry CS (2018) Highly sensitive detection of Salmonella typhimurium using a colorimetric paper-based analytical device coupled with immunomagnetic separation. Anal Chem 90:1035–1043
44. Channon RB, Yang Y, Feibelman KM, Geiss BJ, Dandy DS, Henry CS (2018) Sensitive virus particle detection using an electrochemical paper-based analytical device. ACS Sens. Submitted
45. Hong S, Kwak R, Kim W (2016) Paper-based flow fractionation system applicable to preconcentration and field-flow separation. Anal Chem 88:1682–1687
46. Gong MM, Nosrati R, San Gabriel MC, Zini A, Sinton D (2015) Direct DNA analysis with paper-based ion concentration polarization. J Am Chem Soc 137:13913–13919
47. Ma B, Xie S-F, Liu L, Fang F, Wu Z-Y (2017) Two orders of magnitude electrokinetic stacking of proteins in one minute on a simple paper fluidic channel. Anal Methods 9:2703–2709
48. Jayawardane BM, Coo Ld, Cattrall RW, Kolev SD (2013) The use of a polymer inclusion membrane in a paper-based sensor for the selective determination of Cu(II). Anal Chim Acta 803:106–112
49. Zhu Y, Zhang L, Yang L (2015) Designing of the functional paper-based surface-enhanced Raman spectroscopy substrates for colorants detection. Mater Res Bull 63:199–204
50. Adkins J, Boehle K, Henry C (2015) Electrochemical paper-based microfluidic devices. Electrophoresis 36:1811–1824
51. Mettakoonpitak J, Boehle K, Nantaphol S, Teengam P, Adkins JA, Srisa-Art M, Henry CS (2016) Electrochemistry on paper-based analytical devices. Electroanalysis 28:1420–1436

52. Cate DM, Dungchai W, Cunningham JC, Volckens J, Henry CS (2013) Simple, distance-based measurement for paper analytical devices. Lab Chip 13:2397–2404
53. Pratiwi R, Nguyen MP, Ibrahim S, Yoshioka N, Henry CS, Tjahjono DH (2017) A selective distance-based paper analytical device for copper(II) determination using a porphyrin derivative. Talanta 174:493–499
54. Gomes HI, Sales MG (2015) Development of paper-based color test-strip for drug detection in aquatic environment: Application to oxytetracycline. Biosens Bioelectron 65:54–61
55. Silva TI, Moreira FT, Truta LA, Sales MG (2012) Novel optical PVC probes for on-site detection/determination of fluoroquinolones in a solid/liquid interface: application to the determination of Norfloxacin in aquaculture water. Biosens Bioelectron 36:199–206
56. Li X, Scida K, Crooks RM (2015) Detection of hepatitis B virus DNA with a paper electrochemical sensor. Anal Chem 87:9009–9015
57. Ratnarathorn N, Chailapakul O, Henry CS, Dungchai W (2012) Simple silver nanoparticle colorimetric sensing for copper by paper-based devices. Talanta 99:552–557
58. Chen GH, Chen WY, Yen YC, Wang CW, Chang HT, Chen CF (2014) Detection of mercury(II) ions using colorimetric gold nanoparticles on paper-based analytical devices. Anal Chem 86:6843–6849
59. Dungchai W, Sameenoi Y, Chailapakul O, Volckens J, Henry CS (2013) Determination of aerosol oxidative activity using silver nanoparticle aggregation on paper-based analytical devices. Analyst 138:6766–6773
60. Piyanan T, Athipornchai A, Henry CS, Sameenoi Y (2018) An instrument-free detection of antioxidant activity using paper-based analytical devices coated with nanoceria. Anal Sci 34:97–102
61. Teengam P, Siangproh W, Tuantranont A, Vilaivan T, Chailapakul O, Henry CS (2017) Multiplex paper-based colorimetric DNA sensor using pyrrolidinyl peptide nucleic acid-induced AgNPs aggregation for detecting MERS-CoV, MTB, and HPV oligonucleotides. Anal Chem 89:5428–5435
62. Hu J, Wang L, Li F, Han YL, Lin M, Lu TJ, Xu F (2013) Oligonucleotide-linked gold nanoparticle aggregates for enhanced sensitivity in lateral flow assays. Lab Chip 13:4352–4357
63. Wang Y, Ping J, Ye Z, Wu J, Ying Y (2013) Impedimetric immunosensor based on gold nanoparticles modified graphene paper for label-free detection of Escherichia coli O157:H7. Biosens Bioelectron 49:492–498
64. Kumar A, Hens A, Arun RK, Chatterjee M, Mahato K, Layek K, Chanda N (2015) A paper based microfluidic device for easy detection of uric acid using positively charged gold nanoparticles. Analyst 140:1817–1821
65. Evans E, Gabriel EF, Benavidez TE, Tomazelli Coltro WK, Garcia CD (2014) Modification of microfluidic paper-based devices with silica nanoparticles. Analyst 139:5560–5567
66. Silva TG, de Araujo WR, Munoz RA, Richter EM, Santana MH, Coltro WK, Paixao TR (2016) Simple and sensitive paper-based device coupling electrochemical sample pretreatment and colorimetric detection. Anal Chem 88:5145–5151
67. Pesenti A, Taudte RV, McCord B, Doble P, Roux C, Blanes L (2014) Coupling paper-based microfluidics and lab on a chip technologies for confirmatory analysis of trinitro aromatic explosives. Anal Chem 86:4707–4714
68. Peters KL, Corbin I, Kaufman LM, Zreibe K, Blanes L, McCord BR (2015) Simultaneous colorimetric detection of improvised explosive compounds using microfluidic paper-based analytical devices (μPADs). Anal Methods 7:63–70
69. Musile G, Wang L, Bottoms J, Tagliaro F, McCord B (2015) The development of paper microfluidic devices for presumptive drug detection. Anal Methods 7:8025–8033
70. Cardoso TMG, Channon RB, Adkins JA, Talhavini M, Coltro WKT, Henry CS (2017) A paper-based colorimetric spot test for the identification of adulterated whiskeys. Chem Commun (Camb) 53:7957–7960
71. Liu S, Su W, Ding X (2016) A review on microfluidic paper-based analytical devices for glucose detection. Sensors (Basel) 16(12)

72. Gabriel EF, Garcia PT, Cardoso TM, Lopes FM, Martins FT, Coltro WK (2016) Highly sensitive colorimetric detection of glucose and uric acid in biological fluids using chitosan-modified paper microfluidic devices. Analyst 141:4749–4756
73. Chen X, Chen J, Wang F, Xiang X, Luo M, Ji X, He Z (2012) Determination of glucose and uric acid with bienzyme colorimetry on microfluidic paper-based analysis devices. Biosens Bioelectron 35:363–368
74. Talalak K, Noiphung J, Songjaroen T, Chailapakul O, Laiwattanapaisal W (2015) A facile low-cost enzymatic paper-based assay for the determination of urine creatinine. Talanta 144:915–921
75. Robinson R, Wong L, Monnat R, Fu E (2016) Development of a whole blood paper-based device for phenylalanine detection in the context of PKU therapy monitoring. Micromachines 7:28
76. Dungchai W, Chailapakul O, Henry CS (2010) Use of multiple colorimetric indicators for paper-based microfluidic devices. Anal Chim Acta 674:227–233
77. Sicard C, Glen C, Aubie B, Wallace D, Jahanshahi-Anbuhi S, Pennings K, Daigger GT, Pelton R, Brennan JD, Filipe CD (2015) Tools for water quality monitoring and mapping using paper-based sensors and cell phones. Water Res 70:360–369
78. Zakir Hossain SM, Luckham RE, McFadden MJ, Brennan JD (2009) Reagentless bidirectional lateral flow bioactive paper sensors for detection of pesticides in beverage and food samples. Anal Chem 81:9055–9064
79. Evans E, Gabriel EF, Coltro WK, Garcia CD (2014) Rational selection of substrates to improve color intensity and uniformity on microfluidic paper-based analytical devices. Analyst 139:2127–2132
80. Nery EW, Kubota LT (2016) Evaluation of enzyme immobilization methods for paper-based devices—a glucose oxidase study. J Pharm Biomed Anal 117:551–559
81. Jokerst JC, Adkins JA, Bisha B, Mentele MM, Goodridge LD, Henry CS (2012) Development of a paper-based analytical device for colorimetric detection of select foodborne pathogen. Anal Chem 84:2900–2907
82. Bisha B, Adkins JA, Jokerst JC, Chandler JC, Perez-Mendez A, Coleman SM, Sbodio AO, Suslow TV, Danyluk MD, Henry CS, Goodridge LD (2014) Colorimetric paper-based detection of Escherichia coli, Salmonella spp., and Listeria monocytogenes from large volumes of agricultural water. J Vis Exp (88)
83. Adkins JA, Boehle K, Friend C, Chamberlain B, Bisha B, Henry CS (2017) Colorimetric and electrochemical bacteria detection using printed paper- and transparency-based analytic devices. Anal Chem 89:3613–3621
84. Boehle KE, Gilliand J, Wheeldon CR, Holder A, Adkins JA, Geiss BJ, Ryan EP, Henry CS (2017) Utilizing paper-based devices for antimicrobial-resistant bacteria detection. Angew Chem Int Ed Engl 56:6886–6890
85. Pollock NR, Rolland JP, Kumar S, Beattie PD, Jain S, Noubary F, Wong VL, Pohlmann RA, Ryan US, Whitesides GM (2012) A paper-based multiplexed transaminase test for low-cost, point-of-care liver function testing. Sci Transl Med 4:152ra129
86. Kannan B, Jahanshahi-Anbuhi S, Pelton RH, Li Y, Filipe CD, Brennan JD (2015) Printed paper sensors for serum lactate dehydrogenase using pullulan-based inks to immobilize reagents. Anal Chem 87:9288–9293
87. Nosrati R, Gong MM, San Gabriel MC, Pedraza CE, Zini A, Sinton D (2016) Paper-based quantification of male fertility potential. Clin Chem 62:458–465
88. Yen TH, Chen KH, Hsu MY, Fan ST, Huang YF, Chang CL, Wang YP, Cheng CM (2015) Reprint of 'Evaluating organophosphate poisoning in human serum with paper'. Talanta 145:66–72
89. Posthuma-Trumpie GA, Korf J, van Amerongen A (2009) Lateral flow (immuno)assay: its strengths, weaknesses, opportunities and threats. A literature survey. Anal Bioanal Chem 393:569–582
90. Cheng CM, Martinez AW, Gong J, Mace CR, Phillips ST, Carrilho E, Mirica KA, Whitesides GM (2010) Paper-based ELISA. Angew Chem Int Ed Engl 49:4771–4774

91. Wang S, Ge L, Song X, Yu J, Ge S, Huang J, Zeng F (2012) Paper-based chemiluminescence ELISA: lab-on-paper based on chitosan modified paper device and wax-screen-printing. Biosens Bioelectron 31:212–218
92. Murdock RC, Shen L, Griffin DK, Kelley-Loughnane N, Papautsky I, Hagen JA (2013) Optimization of a paper-based ELISA for human performance biomarker. Anal Chem 85:11634–11642
93. Sun X, Li B, Tian C, Yu F, Zhou N, Zhan Y, Chen L (2018) Rotational paper-based electrochemiluminescence immunodevices for sensitive and multiplexed detection of cancer biomarkers. Anal Chim Acta 1007:33–39
94. Mu X, Zhang L, Chang S, Cui W, Zheng Z (2014) Multiplex microfluidic paper-based immunoassay for the diagnosis of hepatitis C virus infection. Anal Chem 86:5338–5344
95. Khan MS, Pande T, van de Ven TG (2015) Qualitative and quantitative detection of T7 bacteriophages using paper based sandwich ELISA. Colloids Surf B Biointerfaces 132:264–270
96. Shih CM, Chang CL, Hsu MY, Lin JY, Kuan CM, Wang HK, Huang CT, Chung MC, Huang KC, Hsu CE, Wang CY, Shen YC, Cheng CM (2015) Paper-based ELISA to rapidly detect Escherichia coli. Talanta 145:2–5
97. Ramachandran S, Fu E, Lutz B, Yager P (2014) Long-term dry storage of an enzyme-based reagent system for ELISA in point-of-care devices. Analyst 139:1456–1462
98. Schochetman G, Ou C, Jones WK (1988) Polymerase chain reaction. J Infect Dis 158:1154–1157
99. Lafleur LK, Bishop JD, Heiniger EK, Gallagher RP, Wheeler MD, Kauffman P, Zhang X, Kline EC, Buser JR, Kumar S, Byrnes SA, Vermeulen NM, Scarr NK, Belousov Y, Mahoney W, Toley BJ, Ladd PD, Lutz BR, Yager P (2016) A rapid, instrument-free, sample-to-result nucleic acid amplification test. Lab Chip 16:3777–3787
100. Rodriguez NM, Linnes JC, Fan A, Ellenson CK, Pollock NR, Klapperich CM (2015) Paper-based RNA extraction, in situ isothermal amplification, and lateral flow detection for low-cost, rapid diagnosis of Influenza A (H1N1) from clinical specimens. Anal Chem 87:7872–7879
101. Rodriguez NM, Wong WS, Liu L, Dewar R, Klapperich CM (2016) A fully integrated paperfluidic molecular diagnostic chip for the extraction, amplification, and detection of nucleic acids from clinical samples. Lab Chip 16:753–763
102. Rohrman BA, Richards-Kortum RR (2012) A paper and plastic device for performing recombinase polymerase amplification of HIV DNA. Lab Chip 12:3082–3088
103. Huang S, Do J, Mahalanabis M, Fan A, Zhao L, Jepeal L, Singh SK, Klapperich CM (2013) Low cost extraction and isothermal amplification of DNA for infectious diarrhea diagnosis. PLoS One 8:e60059
104. Linnes JC, Rodriguez NM, Liu L, Klapperich CM (2016) Polyethersulfone improves isothermal nucleic acid amplification compared to current paper-based diagnostics. Biomed Microdevices 18:30
105. Teengam P, Siangproh W, Tuantranont A, Henry CS, Vilaivan T, Chailapakul O (2017) Electrochemical paper-based peptide nucleic acid biosensor for detecting human papillomavirus. Anal Chim Acta 952:32–40
106. Yildiz UH, Alagappan P, Liedberg B (2013) Naked eye detection of lung cancer associated miRNA by paper based biosensing platform. Anal Chem 85:820–824
107. Noor MO, Krull UJ (2014) Camera-based ratiometric fluorescence transduction of nucleic acid hybridization with reagentless signal amplification on a paper-based platform using immobilized quantum dots as donors. Anal Chem 86:10331–10339
108. Capitán-Vallvey LF, López-Ruiz N, Martínez-Olmos A, Erenas MM, Palma AJ (2015) Recent developments in computer vision-based analytical chemistry: A tutorial review. Anal Chim Acta 899:23–56
109. Lopez-Ruiz N, Curto VF, Erenas MM, Benito-Lopez F, Diamond D, Palma AJ, Capitan-Vallvey LF (2014) Smartphone-based simultaneous pH and nitrite colorimetric determination for paper microfluidic devices. Anal Chem 86:9554–9562

110. Shen L, Hagen JA, Papautsky I (2012) Point-of-care colorimetric detection with a smartphone. Lab Chip 12:4240–4243
111. Martinez AW, Phillips ST, Carrilho E, Thomas SW, Sindi H, Whitesides GM (2008) Simple telemedicine for developing regions: Camera phones and paper-based microfluidic devices for real-time, off-site diagnosis. Anal Chem 80:3699–3707
112. Klasner SA, Price AK, Hoeman KW, Wilson RS, Bell KJ, Culbertson CT (2010) Paper-based microfluidic devices for analysis of clinically relevant analytes present in urine and saliva. Anal Bioanal Chem 397:1821–1829
113. Dungchai W, Chailapakul O, Henry CS (2009) Electrochemical detection for paper-based microfluidics. Anal Chem 81:5821–5826
114. Arduini F, Micheli L, Moscone D, Palleschi G, Piermarini S, Ricci F, Volpe G (2016) Electrochemical biosensors based on nanomodified screen-printed electrodes: Recent applications in clinical analysis. TrAC Trends Anal Chem 79:114–126
115. Nie Z, Nijhuis CA, Gong J, Chen X, Kumachev A, Martinez AW, Narovlyansky M, Whitesides GM (2010) Electrochemical sensing in paper-based microfluidic devices. Lab Chip 10:477–483
116. Maattanen A, Ihalainen P, Pulkkinen P, Wang SX, Tenhu H, Peltonen J (2012) Inkjet-printed gold electrodes on paper: characterization and functionalization. ACS Appl Mater Interfaces 4:955–964
117. Hu CG, Bai XY, Wang YK, Jin W, Zhang X, Hu SS (2012) Inkjet printing of nanoporous gold electrode arrays on cellulose membranes for high-sensitive paper-like electrochemical oxygen sensors using ionic liquid electrolytes. Anal Chem 84:3745–3750
118. Dossi N, Toniolo R, Piccin E, Susmel S, Pizzariello A, Bontempelli G (2013) Pencil-drawn dual electrode detectors to discriminate between analytes comigrating on paper-based fluidic devices but undergoing electrochemical processes with different reversibility. Electroanalysis 25:2515–2522
119. Dossi N, Toniolo R, Pizzariello A, Impellizzieri F, Piccin E, Bontempelli G (2013) Pencil-drawn paper supported electrodes as simple electrochemical detectors for paper-based fluidic devices. Electrophoresis 34:2085–2091
120. Santhiago M, Kubota LT (2013) A new approach for paper-based analytical devices with electrochemical detection based on graphite pencil electrodes. Sens Actuat B Chem 177:224–230
121. Dossi N, Toniolo R, Impellizzieri F, Bontempelli G (2014) Doped pencil leads for drawing modified electrodes on paper-based electrochemical devices. J Electroanal Chem 722:90–94
122. Fosdick SE, Anderson MJ, Renault C, DeGregory PR, Loussaert JA, Crooks RM (2014) Wire, mesh, and fiber electrodes for paper-based electroanalytical devices. Anal Chem 86:3659–3666
123. Adkins JA, Henry CS (2015) Electrochemical detection in paper-based analytical devices using microwire electrodes. Anal Chim Acta 891:247–254
124. Shiroma LY, Santhiago M, Gobbi AL, Kubota LT (2012) Separation and electrochemical detection of paracetamol and 4-aminophenol in a paper-based microfluidic device. Anal Chim Acta 725:44–50
125. Siegel AC, Phillips ST, Wiley BJ, Whitesides GM (2009) Thin, lightweight, foldable thermochromic displays on paper. Lab Chip 9:2775–2781
126. Carvalhal RF, Kfouri MS, Piazetta MHD, Gobbi AL, Kubota LT (2010) Electrochemical detection in a paper-based separation device. Anal Chem 82:1162–1165
127. Lankelma J, Nie ZH, Carrilho E, Whitesides GM (2012) Paper-based analytical device for electrochemical flow-injection analysis of glucose in urine. Anal Chem 84:4147–4152
128. Jagadeesan KK, Kumar S, Sumana G (2012) Application of conducting paper for selective detection of troponin. Electrochem Commun 20:71–74
129. Ge SG, Zhang LN, Zhang Y, Liu HY, Huang JD, Yan M, Yu JH (2015) Electrochemical K-562 cells sensor based on origami paper device for point-of-care testing. Talanta 145:12–19

130. Ge SG, Zhang Y, Zhang L, Liang LL, Liu HY, Yan M, Huang JD, Yu JH (2015) Ultrasensitive electrochemical cancer cells sensor based on trimetallic dendritic Au@PtPd nanoparticles for signal amplification on lab-on-paper device. Sens Actuat B Chem 220:665–672
131. Cunningham JC, Brenes NJ, Crooks RM (2014) Paper electrochemical device for detection of DNA and thrombin by target-induced conformational switching. Anal Chem 86:6166–6170
132. Ge L, Wang SM, Yu JH, Li NQ, Ge SG, Yan M (2013) Molecularly imprinted polymer grafted porous Au-paper electrode for an microfluidic electro-analytical origami device. Adv Funct Mater 23:3115–3123
133. Wang PP, Sun GQ, Ge L, Ge SG, Yu JH, Yan M (2013) Photoelectrochemical lab-on-paper device based on molecularly imprinted polymer and porous Au-paper electrode. Analyst 138:4802–4811
134. Jia W, Bandodkar AJ, Valdés-Ramírez G, Windmiller JR, Yang Z, Ramírez J, Chan G, Wang J (2013) Electrochemical tattoo biosensors for real-time noninvasive lactate monitoring in human perspiration. Anal Chem 85:6553–6560
135. Cinti S, Arduini F, Carbone M, Sansone L, Cacciotti I, Moscone D, Palleschi G (2015) Screen-printed electrodes modified with carbon nanomaterials: a comparison among carbon black, carbon nanotubes and graphene. Electroanalysis 27:2230–2238
136. Ruecha N, Rangkupan R, Rodthongkum N, Chailapakul O (2014) Novel paper-based cholesterol biosensor using graphene/polyvinylpyrrolidone/polyaniline nanocomposite. Biosens Bioelectron 52:13–19
137. Arduini F, Forchielli M, Amine A, Neagu D, Cacciotti I, Nanni F, Moscone D, Palleschi G (2015) Screen-printed biosensor modified with carbon black nanoparticles for the determination of paraoxon based on the inhibition of butyrylcholinesterase. Microchim Acta 182:643–651
138. Michopoulos A, Kouloumpis A, Gournis D, Prodromidis MI (2014) Performance of layer-by-layer deposited low dimensional building blocks of graphene-Prussian blue onto graphite screen-printed electrodes as sensors for hydrogen peroxide. Electrochim Acta 146:477–484
139. Kong FY, Gu SX, Li WW, Chen TT, Xu Q, Wang W (2014) A paper disk equipped with graphene/polyaniline/Au nanoparticles/glucose oxidase biocomposite modified screen-printed electrode: Toward whole blood glucose determination. Biosens Bioelectron 56:77–82
140. Liana DD, Raguse B, Wieczorek L, Baxter GR, Chuah K, Gooding JJ, Chow E (2013) Sintered gold nanoparticles as an electrode material for paper-based electrochemical sensors. RSC Adv 3:8683–8691
141. Pedroza RHP, Serrano N, Díaz-Cruz JM, Ariño C, Esteban M (2016) Integration of commercial screen-printed electrodes into a voltammetric electronic tongue for the analysis of aminothiols. Electroanalysis 28:1570–1577
142. Apilux A, Dungchai W, Siangproh W, Praphairaksit N, Henry CS, Chailapakul O (2010) Lab-on-paper with dual electrochemical/colorimetric detection for simultaneous determination of gold and iron. Anal Chem 82:1727–1732
143. Ruecha N, Rodthongkum N, Cate DM, Volckens J, Chailapakul O, Henry CS (2015) Sensitive electrochemical sensor using a graphene-polyaniline nanocomposite for simultaneous detection of Zn(II), Cd(II), and Pb(II). Anal Chim Acta 874:40–48
144. Noiphung J, Songjaroen T, Dungchai W, Henry CS, Chailapakul O, Laiwattanapaisal W (2013) Electrochemical detection of glucose from whole blood using paper-based microfluidic devices. Anal Chim Acta 788:39–45
145. Lisak G, Cui J, Bobącka J (2015) Paper-based microfluidic sampling for potentiometric determination of ions. Sens Actuators B Chim 207:933–939
146. Lan WJ, Zou XU, Hamedi MM, Hu JB, Parolo C, Maxwell EJ, Buhlmann P, Whitesides GM (2014) Paper-based potentiometric ion sensing. Anal Chem 86:9548–9553
147. Wang Y, Liu H, Wang P, Yu J, Ge S, Yan M (2015) Chemiluminescence excited photoelectrochemical competitive immunosensing lab-on-paper device using an integrated paper supercapacitor for signal amplication. Sens Actuators B Chim 208:546–553

148. Wu L, Ma C, Zheng X, Liu H, Yu J (2015) Paper-based electrochemiluminescence origami device for protein detection using assembled cascade DNA–carbon dots nanotags based on rolling circle amplication. Biosens Bioelectron 68:413–420
149. Zhang Y, Zuo P, Ye B-C (2015) A low-cost and simple paper-based microfluidic device for simultaneous multiplex determination of different types of chemical contaminants in food. Biosens Bioelectron 68:14–19
150. Qu L-L, Li D-W, Xue J-Q, Zhai W-L, Fossey JS, Long Y-T (2012) Batch fabrication of disposable screen printed SERS arrays. Lab Chip 12:876–881
151. Wang H, Liu J, Cooks RG, Ouyang Z (2010) Paper spray for direct analysis of complex mixtures using mass spectrometry. Angew Chem Int Ed 49:877–880
152. Hu J, Wang S, Wang L, Li F, Pingguan-Murphy B, Lu TJ, Xu F (2014) Advances in paper-based point-of-care diagnostics. Biosens Bioelectron 54:585–597
153. Ellerbee AK, Phillips ST, Siegel AC, Mirica KA, Martinez AW, Striehl P, Jain N, Prentiss M, Whitesides GM (2009) Quantifying colorimetric assays in paper-based microfluidic devices by measuring the transmission of light through paper. Anal Chem 81:8447–8452
154. Lee D-S, Jeon BG, Ihm C, Park J-K, Jung MY (2011) Size limits the formation of liquid jets during bubble bursting. Lab Chip 11:120–126
155. Nie Z, Deiss F, Liu X, Akbulut O, Whitesides GM (2010) Integration of paper-based microfluidic devices with commercial electrochemical readers. Lab Chip 10:3163–3169
156. Zhang L, Yang W, Yang Y, Liu H, Gu Z (2015) Smartphone-based point-of-care testing of salivary [small alpha]-amylase for personal psychological measurement. Analyst 140:7399–7406
157. Elizalde E, Urteaga R, Berli CLA (2015) Rational design of capillary-driven flows for paper-based microfluidics. Lab Chip 15:2173–2180
158. Toley BJ, Wang JA, Gupta M, Buser JR, Lafleur LK, Lutz BR, Fu E, Yager P (2015) A versatile valving toolkit for automating fluidic operations in paper microfluidic devices. Lab Chip 15:1432–1444
159. Cunningham JC, Scida K, Kogan MR, Wang B, Ellington AD, Crooks RM (2015) Paper diagnostic device for quantitative electrochemical detection of ricin at picomolar levels. Lab Chip 15:3707–3715
160. Adkins JA, Noviana E, Henry CS (2016) Development of a quasi-steady flow electrochemical paper-based analytical device. Anal Chem 88:10639–10647
161. Chandra Sekar N, Mousavi Shaegh SA, Ng SH, Ge L, Tan SN (2014) A paper-based amperometric glucose biosensor developed with Prussian blue-modified screen-printed electrodes. Sens Actuat B Chem 204:414–420
162. Mudanyali O, Dimitrov S, Sikora U, Padmanabhan S, Navruz I, Ozcan A (2012) Integrated rapid-diagnostic-test reader platform on a cellphone. Lab Chip 12:2678–2686
163. Oncescu V, O'Dell D, Erickson D (2013) Smartphone based health accessory for colorimetric detection of biomarkers in sweat and saliva. Lab Chip 13:3232–3238
164. Salles MO, Meloni GN, de Araujo WR, Paixao TRLC (2014) Explosive colorimetric discrimination using a smartphone, paper device and chemometrical approach. Anal Methods 6:2047–2052

Chapter 4
How Are These Devices Manufactured?

Giorgio Gianini Morbioli, Thiago Mazzu-Nascimento, Amanda M. Stockton, and Emanuel Carrilho

Abstract Paper-based devices are a portable and low-cost technology alternative among conventional analytical tools for point-of-care testing that have been growing in popularity due to their versatility and ease of use. Since their introduction in 2007, different methods to fabricate such devices have been proposed in literature, including photolithography, plotting, inkjet printing, wax printing, and stamping, among other methods, each with its unique advantages and drawbacks. Here, we present the most common current fabrication methods of microfluidic paper-based analytical devices (μPADs) in two and three dimensions, comparing their processes of fabrication, resolution of the features, and costs associated with their manufacture presenting insights towards the most adequate choice of fabrication method.

4.1 Design Criteria

Independent of the patterning method chosen to manufacture paper-based microchips, the design of paper-based microfluidic analytical devices (μPADs) plays a major role in their performance. In order to maintain the advantages of portability and low cost [1–3] microfluidic structures are designed with a small footprint. However, there are physical limits for the miniaturization of such structures. If their hydrophobic barriers or the hydrophilic channels do not have enough resolution (sharpness of the boundaries hydrophilic/hydrophobic), then the functionality of the device is compromised [4]. The design of each section of a device must be designed rationally in a μPAD to avoid the introduction of bias and to maximize chip capabilities [5].

G. G. Morbioli · A. M. Stockton
School of Chemistry and Biochemistry, Georgia Institute of Technology, Atlanta, GA, USA

T. Mazzu-Nascimento · E. Carrilho (✉)
Instituto de Química de São Carlos, Universidade de São Paulo, São Carlos, SP, Brazil

Instituto Nacional de Ciência e Tecnologia de Bioanalítica, Campinas, SP, Brazil
e-mail: emanuel@iqsc.usp.br

K. J. Land (ed.), *Paper-based Diagnostics*,
https://doi.org/10.1007/978-3-319-96870-4_4

The microchip architecture of both 2- and 3D structures can be divided into zones: (1) sample introduction, (2) fluidic manipulation, and (3) detection zone. We discuss each zone in detail in this section of the chapter.

4.1.1 Sample Introduction

Common practice in the paper-based microfluidics area involves the insertion of samples into microdevices using pipettes, which enhances the figures of merit of the analytical method, especially in the cases of semiquantitative and quantitative approaches. However, these devices are envisioned to be used as point-of-care testing (POCT) tools by people with little to no training, ideally without the use of external equipment, according to the World Health Organization (WHO) in the ASSURED Challenge (*A*ffordable, *S*ensitive, *S*pecific, *U*ser-friendly *R*apid and robust, *E*quipment-free, and *D*eliverable) [6]. Even though some low-cost alternatives to the use of micropipettes have been proposed [7], an equipment-free approach to introduce the sample into the μPAD is the preferred method in products developed for the end user, and this should be taken into account when designing μPADs.

Since manufacturing methods produce consistent μPADs, with similar specifications, it is plausible that these μPADs present the same hydrophilic area and that the sample volume wicked by these devices—in the absence of evaporation—will be the same, potentially eliminating the need for micropipettes. This concept was first presented and explored by Martinez et al. [7], who observed that their 2D-μPADs wicked sample volumes (~5 μL) in a reproducible manner. There is a trade between using a pipette to accurately measure the volume of sample being used and the accuracy of the measurement: the authors reported that the error in their concentration determinations was smaller when using a micropipette to introduce the sample into the device, in comparison with larger errors when using just the device itself to absorb a fixed and precise volume of sample [7].

To make use of this approach, it is necessary to bring the device into contact with the sample, making it useful to introduce a hydrophilic path starting from device extremities (Fig. 4.1a). To avoid interference with fluidic paths, reaction zones, or detection spots (Fig. 4.1b) during handling, it is also useful to have space in the device for manual manipulation (Fig. 4.1c).

4.1.2 Fluidic Manipulation

Once introduced into the device, the sample permeates the microdevice through the hydrophilic channels due to capillarity action. Capillary-driven flow depends on both the substrate type (porosity, mineral filling charge, additives [8]) and the geometrical shape of the channels [9], making rational design critical. Additionally,

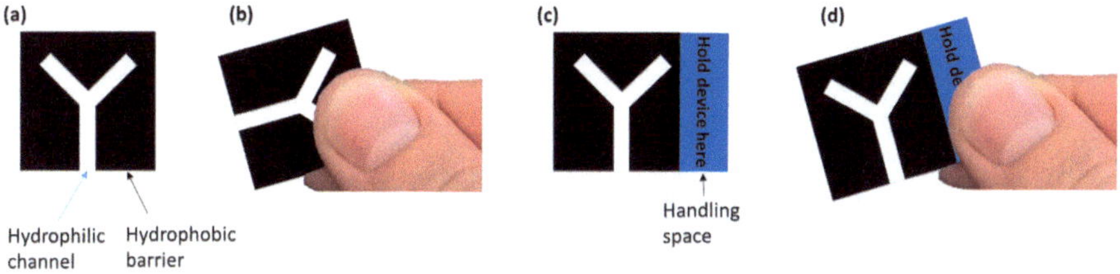

Fig. 4.1 Factors that should be considered when designing the sample introduction method in a μPAD. (**a**) Hydrophilic channel placed at one extremity of the paper-based device. (**b**) A μPAD being handled incorrectly due to lack of space for manipulation. (**c**) A μPAD with a designated space for correct handling and labeling of samples. (**d**) A μPAD with a designated space for handling being manipulated correctly

different unit operations can be embedded on paper-based devices, including filtration [7], pre-concentration [10, 11], multiplexing [12], and mixing [13], depending on the application of the device. The integration of multiple tools in a single, small-footprint device relies on rational microdevice design.

4.1.2.1 Sample Distribution and Multiplexing

Sample distribution in the microfluidic device depends on the fluidic resistance of the hydrophilic paths. However, for design, it would be more useful to predict a channel that will provide a desired fluid velocity rather than empirically determining fluid velocity in device design iterations, and from the geometrical shape of the hydrophilic channels it is possible to estimate the capillary flow rate in the structure [9]. The cross-sectional profile of the hydrophilic channel ($A(x)$) as a function of the desired fluid velocity function ($v(x)$) is given by Eq. (4.1) [9]:

$$A(x) = \frac{v_0 A_0}{v(x)} \exp\left[-\frac{1}{D}\int_0^x v\left(x'\right) \mathrm{d}x'\right] \tag{4.1}$$

where $A(x)$ is the cross-sectional profile, $v(x)$ is the specified velocity, v_0A_0 is the initial boundary condition, D is a diffusive coefficient that characterizes the substrate–fluid system, and x is the channel length.

If one desires to design a two-dimensional microfluidic system that maintains constant fluidic velocity throughout the structure, the channel must have an exponential decaying width (Fig. 4.2) [9], according to Eq. (4.2):

$$w(x) = w_0 \exp\left(-\frac{x}{l_0}\right) \tag{4.2}$$

where $w(x)$ is the variable width; w_0 and l_0 are the specifications of the region with uniform width and length, respectively; and x is the distance at the liquid front.

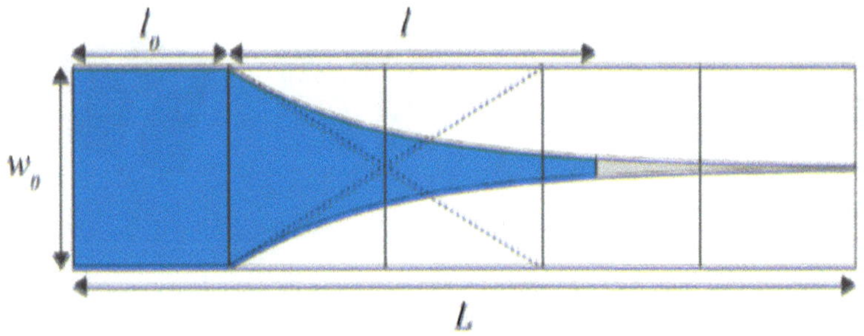

Fig. 4.2 Design of a hydrophilic channel on a sheet with uniform thickness that maintains constant velocity of the liquid front. Reproduced in part from ref. [9] with permission of The Royal Society of Chemistry

Another important design parameter is the fluidic resistance of the hydrophilic channels, especially when dealing with multiplexers. If there is differential resistance between channels, then the sample will permeate preferentially through the path of least resistance, analogously to an electric circuit [14], biasing sample distribution in the device [5]. In order to maintain a small footprint of the device to reduce manufacture and transportation costs, the device should maximize capabilities. This is especially true in three-dimensional devices [5], because repeatability of layer assembling is highly dependent on the number of layers [15] and requires extra steps of labor-intense work [12].

4.1.2.2 Unit Operations

Many basic devices can be fabricated with simple channels and therefore must only meet the requirements covered in the above discussion [16]. However, more elaborate operations are often employed in μPADs to diminish sample preparation, such as timers [17], flow interrupters [18], valving [19], storage [20], and mixers [13].

Timers and valves accomplish a common objective: To allow the flow of fluid through a channel after a desired time, but employ different mechanisms. Valve operation requires the direct intervention of the operator, enabling the hydrophilic contact across a discontinuity in the channel (Fig. 4.3a) [19]. Valves are more frequently employed in exploratory studies, as they enable variation of the flow time in a facile, tunable fashion and fluidic connection can be established or interrupted as many times necessary [19].

The operational principle of a timer is easily seen through the example reported by Lutz et al. [17]. After introduction of a saturated solution of sugar into the hydrophilic channel the solvent is evaporated, leaving a crystalized sugar barrier in the channel [17]. During the assay, when the active fluid meets the barrier, there is solubilization of the sugar, changing the viscosity of the fluid and reducing flow velocity according to Eq. (4.3) [17] (Fig. 4.3b):

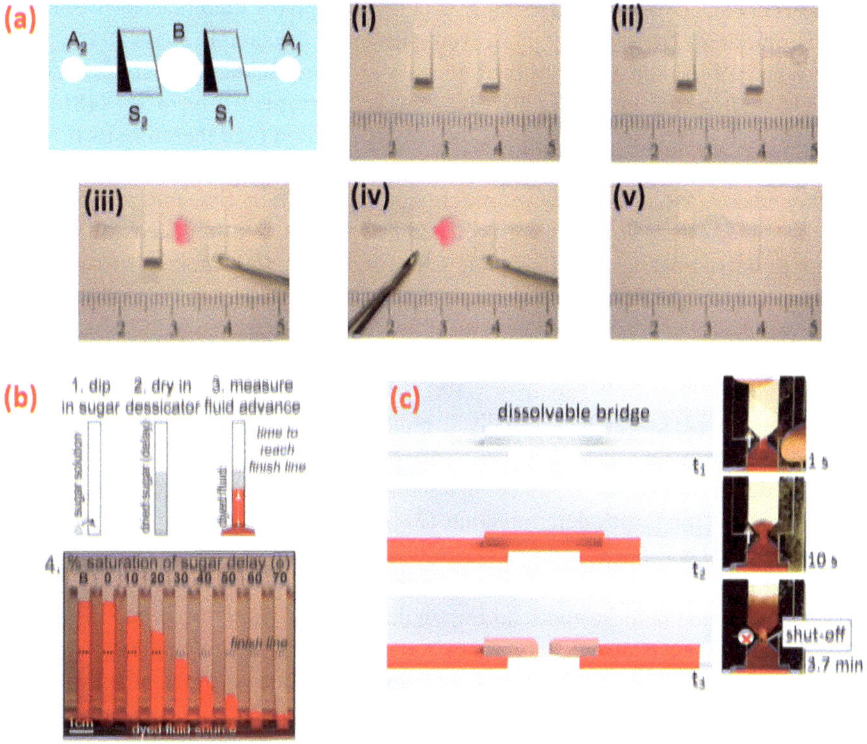

Fig. 4.3 Examples of unit operations performed on paper-based devices. (**a**) Valving system demonstration. A_1 is a reservoir containing HCl solution. A_2 is a reservoir containing NaOH solution. B is a reactor site containing phenolphthalein. S_1 and S_2 are switches operated manually. (i) S_1 and S_2 open, before solution introduction. (ii) S_1 and S_2 open, after solution introduction. (iii) S_1 closed and S_2 open, changing the color of the pH indicator to pink. (iv) S_1 and S_2 closed, allowing for HCl solution to neutralize the basic solution. (v) S_1 and S_2 closed, changing the color of the pH indicator from pink to no color. Reprinted (adapted) with permission from ref. [19]. Copyright (2008) American Chemical Society. (**b**) Illustrations 1. and 2. represent the schematic preparation of the sugar-bridge timer. Illustration 3. represents the schematic operation of the barrier, while the image 4. depicts the operation of the sugar delay barriers containing different amounts of sugar. Reproduced in part from ref. [17] with permission of The Royal Society of Chemistry. (**c**) Representation of the operation of the sugar-bridge flow interrupter. Reprinted (adapted) with permission from ref. [18]. Copyright (2013) American Chemical Society

$$L^2 = \frac{\gamma D t}{4\eta} \tag{4.3}$$

where L is the distance travelled by the fluid, γ is the fluid surface tension, η is the fluid viscosity, D is a diffusive coefficient that characterizes the substrate–fluid system, and t is the time.

Valves can also be based on the dissolution of sugars and used to create a fluidic interrupter. In this implementation, a sugar bridge is created in a microfluidic channel between two hydrophilic paper channels. When the fluid reaches the bridge, the dissolution of the sugar by the solvent present in the sample starts taking place [18]. When all the sugar is dissolved, there is no more fluidic contact between the adjacent hydrophilic channels, and the flow is interrupted (Fig. 4.3c) [18]. This approach also affects the viscosity of the sample solution (and therefore its flow).

Timers and fluidic interrupters require no external intervention of the operator, and are useful for operations in which fluidic operation times have been pre-established. These operations are considered irreversible: once reduced, the velocity cannot be raised again, and the fluidic connection that was interrupted cannot be established again without external manipulation.

To achieve the vision of deployment in remote locations without access to equipment and reagents [20], all necessary assay reagents should ideally be stored in the device. The shelf life of a μPAD depends on the stability of the least stable reagent [21], which is usually a biological component like enzymes that lose activity with time [7]. However, cross-reactions also can take place between reagents depending on the assay being conducted (such as the reagents used in the Griess reagent for nitrite determination [22]), which can impair the use of these assays on μPADs if not accounted for. To overcome this challenge, the creation of reagent storage spots in the microdevice is employed: the reagent components are separately applied in these spots, and when the sample is introduced into the device the solvent from the sample solubilizes the stored reagents, carrying them with the flow (Fig. 4.4) [20, 22]. The reagents are then brought into contact only during the assay development, preventing cross-reaction between them.

Mixing is another unit operation required in multiple processes, promoting contact between reagents and their reaction. However, laminar flow impairs mixing between solutions [13]. For example, when two different flows are brought together in parallel in a channel (Fig. 4.5a), the passive mixing between these solutions is only due to diffusion of species at the laminar flow interface [13]. Passive mixing mechanisms require long channels to promote the mixing of these solutions because of the small contact area between solutions, significantly increasing the footprint of the device and the time required for the assay to complete [13].

If passive mixing is chosen for the μPAD then it becomes necessary to increase the contact area between the adjacent flows. The flat Y-mixer developed by Osborn and co-workers [13] makes use of this fact by stacking two hydrophilic channels

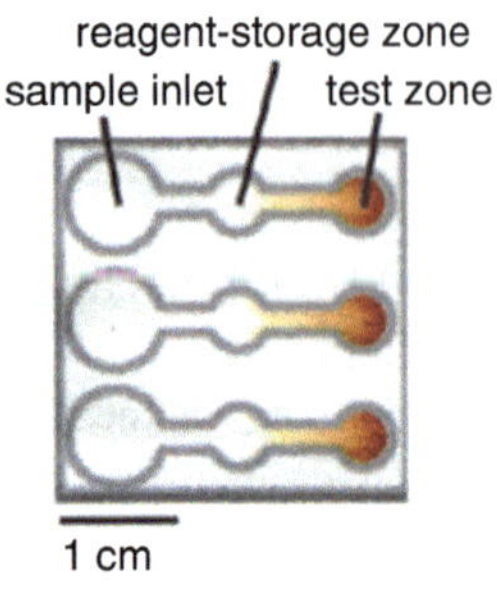

Fig. 4.4 Use of a reagent-storage zone in a paper-based device. Reprinted with permission from ref. [20]. Copyright (2012) American Chemical Society

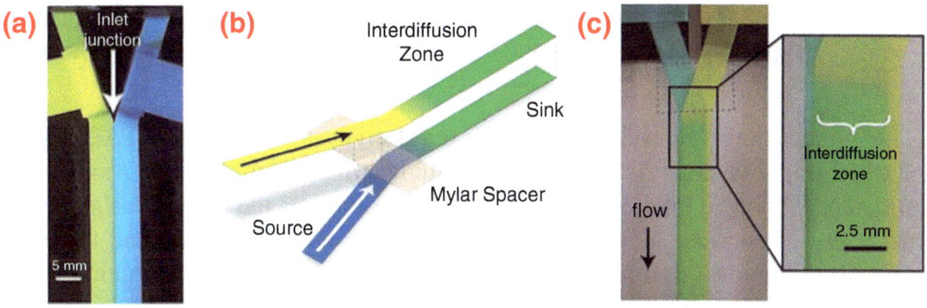

Fig. 4.5 Mixing in paper-based devices. (**a**) Y-device showing the passive mixing of tartrazine (yellow) and erioglaucine (blue) with a small contact area. (**b**) and (**c**) show a flat Y-mixer with a larger contact area. Reproduced (adapted) from ref. [13] with permission of The Royal Society of Chemistry

one on top of another (Fig. 4.5b, c), maximizing the contact area and minimizing the distance required for diffusive mixing [13]. The authors have reported full mixing after a distance of ~2.5 mm, with an enhancement factor in the mixing time of 125,000 over a single-layer Y chip (Fig. 4.5a) (mixing time of 2.5 s for a system with a flow rate of 0.01 cm s^{-1}) [13], demonstrating that passive mixing can be functional in μPADs. As the authors state, this mechanism can be used only in three-dimensional structures, due to the requirement for stacking.

4.1.3 Detection Zone

The detection zone is the destination of a sample being analyzed in a μPAD. The detection zone does more than reporting the result of the assay, as it also compartmentalizes different zones to prevent cross-contamination between spots. In some devices, the detection zone is also the reaction chamber, presenting then a dual function [7]. However, due to the common small footprint of μPADs, a short distance separates neighboring detection zones and depending on the analytes being sensed, reagents for the assays, and by-products of reactions happening in the device, cross-contamination can occur even without fluidic contact between adjacent areas.

Because the previously discussed condition is very system dependent, the design of the detection zone in a μPAD becomes an iterative process: it is necessary to test if any of the components of the assay, including the analyte of interest and reaction by-products, will cause an enhancement in the signal of neighboring detection zones. In a positive case, it is necessary to redesign the microfluidic platform increasing the distance between adjacent zones until no further cross-contamination persists.

4.1.4 Three-Dimensional Microfluidic Paper-Based Analytical Devices: 3D-μPAD

Three-dimensional microfluidic paper-based analytical devices (3D-μPADs) are a stack of single-layer μPADs [5], enabling fluid permeation through the *z*-axis in the structure. This new fluidic motion direction also impacts the device design, being necessary to consider how the fluid is transported between layers, and therefore impacts the rational design of each single layer and the fluid transportation in the *x*–*y* plane [5]. An imbalance in the hydrodynamic resistance in either a layer or between adjacent layers in a μPAD causes preferential paths to liquid permeation, which affects sample distribution across the device and therefore the output on such devices [5].

The layer assembly method of 3D-μPADs also plays a significant role in device functionality. Using a reversible assembling method, such as an external housing unit using screws to keep the paper layers united [23] (Fig. 4.6), may result in a force imbalance to be introduced into the system, favoring fluidic dispersion in some regions of the μPAD [5]. A more uniform and reproducible external layer-assembling method alleviates this issue, such as the use of an external magnetic apparatus that uses two flat stainless steel plates united by magnets [5, 24].

External assembly setups for μPADs require device handling during its introduction into the external apparatus, which can require some user training and the equipment itself, which are the major drawbacks of these methods. An alternative approach is the use of irreversible layer-assembling methods [5] using tape and cellulose powder (Fig. 4.7a) [12], adhesive spray (Fig. 4.7b) [15], or toner and lamination (Fig. 4.7c) [25]. Such alternatives result in ready-to-use devices that do not require user intervention for assembly. However, disadvantages of these methods include extra steps in device manufacture and sometimes a labor-intense work for device assembling [12], translating into more expensive fabrication costs.

4.1.5 General Considerations

Some features are not detrimental for μPAD usage but can improve their functionality and ease their manipulation. For example, the inclusion of a region for writing on the μPAD minimizes patient misidentification, a relevant issue in clinical diagnosis (Fig. 4.8) [20]. This extra space can also improve device manipulation (Fig. 4.1c) but requires a larger device footprint.

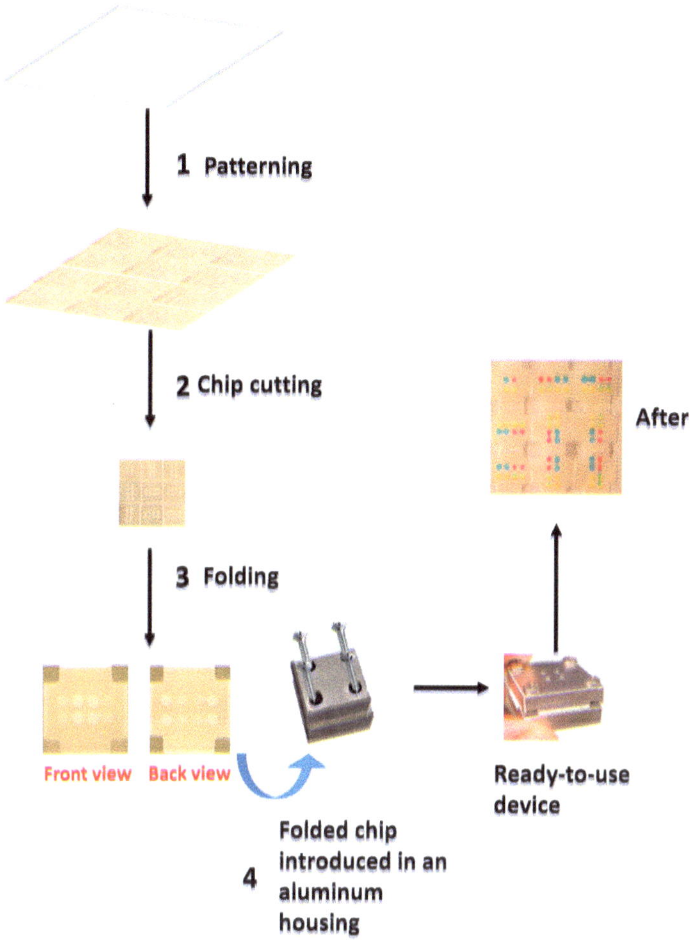

Fig. 4.6 External aluminum housing and screw method to keep μPAD layers together. Adapted with permission from ref. [23]. Copyright (2011) American Chemical Society

Another feature that shows promising results is the full-enclosure μPADs (Fig. 4.8) [20] to protect channels from contamination and mishandling (Fig. 4.1b) and minimize sample losses by evaporation [20] without modifying the device footprint. However, depending on the method used for enclosing the devices, the performance of them can be impaired. For example, when toner printing is used as the enclosing method in μPADs supporting bioassays using enzymes, the heat generated during the printing process can reduce up to 90% of enzymatic activity, indicating that the enclosure method must be compatible with the assay performed on the microdevice [20].

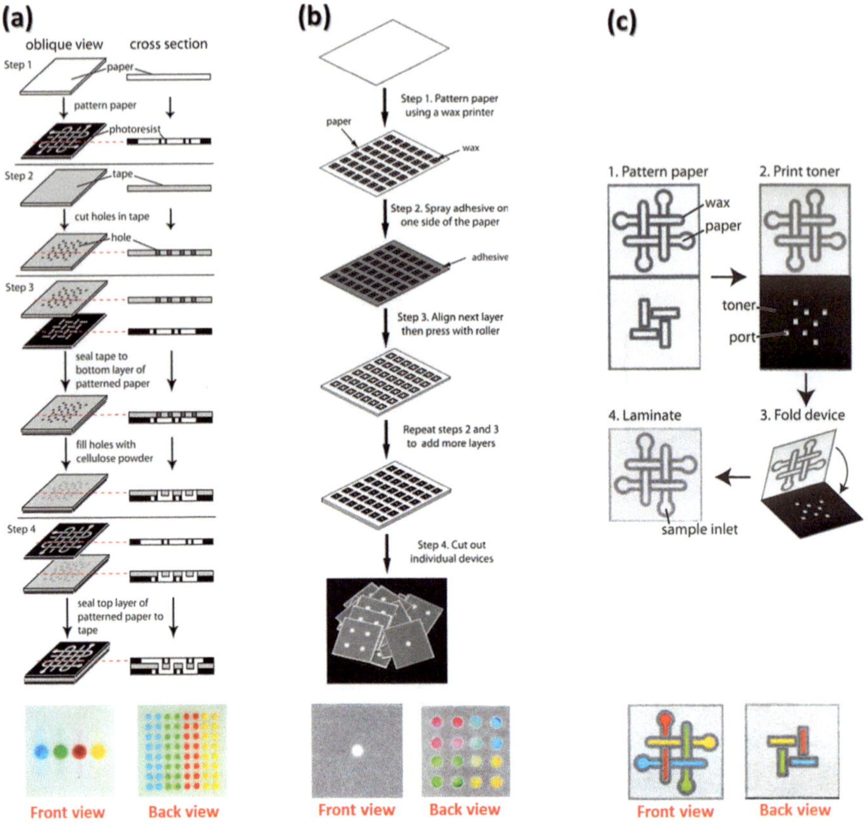

Fig. 4.7 Irreversible layer-assembly methods. (**a**) Tape and cellulose powder method. Adapted with permission from ref. [12]. Copyright (2008) National Academy of Sciences. (**b**) Adhesive spray method. Adapted from ref. [15] with permission of The Royal Society of Chemistry. (**c**) Toner and lamination method. Adapted from ref. [25] with permission of The Royal Society of Chemistry

4.2 Current Methods

The first patterning method for μPAD fabrication was presented in 2007 and used photolithography to create hydrophobic barrier selectivity on the body of the paper, generating hydrophilic channels to control the solvent flow and to create individual reaction zones in the device [2, 26, 27]. Since then, many authors proposed different methods for μPAD fabrication, each one with its considerations. The manufacture method choice depends on both the instrumentation available and the microchip requirements, regarding the resolution of hydrophilic channels and hydrophobic barriers, besides solvent compatibility. Table 4.1 presents the most common fabrication methods for μPAD manufacture [27], discussed in detail in each section.

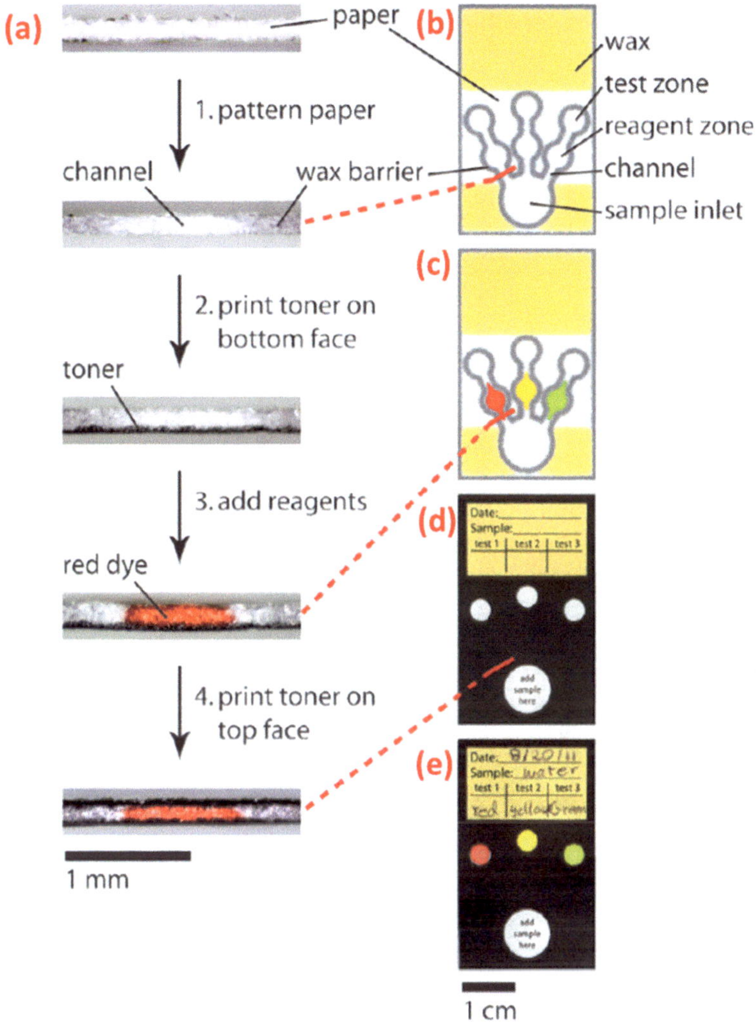

Fig. 4.8 Toner-enclosed μPADs. (**a**) Cross section of an enclosed μPAD during its manufacture. (**b**) Device after the patterning method—wax printing in this case. (**c**) Introduction of dyes before full enclosing with toner. (**d**) Fully enclosed μPAD. (**e**) Identified device after the introduction of 20 μL of water into the sample inlet. Reprinted with permission from ref. [20]. Copyright (2012) American Chemical Society

4.2.1 Photolithography

Photolithography was the first patterning method to manufacture μPADs as we know them today, introduced in 2007 by the Whitesides Group at Harvard University [26]. This seminal work created well-defined hydrophilic areas on paper

Table 4.1 Most common fabrication methods for μPAD manufacture

Fabrication technique	Channel resolution	Barrier resolution	Principle	Approach	Advantages	Disadvantages	Ref.
Photolithography	186 ± 13 μm	248 ± 13 μm	Physical blocking of pores in paper	Entire hydrophobization followed by selective dehydrophobization	High resolution of hydrophilic channels	High-cost equipment; additional washings required	[26]
Plotting	N/A	N/A	Physical blocking of pores in paper	Selective hydrophobization	Low cost of modifying agent; flexible devices	Low resolution of hydrophobic barriers; low production yield	[28]
Plasma treatment	N/A	N/A	Chemical modification of fiber surface	Entire hydrophobization followed by selective dehydrophobization	Low cost of modifying agent	Requires different masks	[19]
Wax printing	561 ± 45 μm	850 ± 50 μm	Physical deposition of reagent on fiber surface	Selective hydrophobization	Low cost of modifying agent; mass production	Wax printer required; requires additional thermal cycles	[4, 29]
Wax dipping	639 ± 7 μm	N/A	Physical deposition of reagent on fiber surface	Selective hydrophobization	Low cost of modifying agent; mass production	Low channel resolution; requires different masks for different designs	[30]
Inkjet printing	272 ± 19 μm	N/A	Chemical modification of fiber surface	Selective hydrophobization	Low cost of modifying agent; mass production	Modified printer required; requires additional thermal cycles	[31]

Inkjet etching	~550 µm	N/A	Physical deposition of reagent on fiber surface	Entire hydrophobization followed by selective dehy-drophobization	Requires only one device to create channels and to deposit reagents	Requires ten printing cycles; customized equipment required; not recommended for mass production	[32]
Flexography printing	500 ± 30 µm	N/A	Physical deposition of reagent on fiber surface	Selective hydrophobization	Can utilize molds from other process; does not require additional thermal cycles	Requires two printing cycles and different molds; quality of devices depends on the paper of choice	[33]
Screen printing	~650 µm	~1300 µm	Physical deposition of reagent on fiber surface	Selective hydrophobization	Simple processing	Low channel resolution; requires different masks for different designs	[34]
Laser treatment	62 ± 1 µm	~250 µm	Physical blocking of pores in paper	Entire hydrophobization followed by selective dehy-drophobization	High resolution of hydrophilic channels	Hydrophilic channels do not permit lateral flow; requires additional coating	[35]
Stamping	2.6 ± 0.1 mm	1.4 ± 0.1 mm	Physical deposition of reagent on fiber surface	Selective hydrophobization	Low cost of modifying agent; mass production	Requires metal mold for each design; requires additional thermal cycles	[36, 37]

N/A Not available

Modified from Li, X.; Ballerini, D. R.; Shen, W. A perspective on paper-based microfluidics: Current status and future trends. **Biomicrofluidics**, v. 6, n. 1, p. 11301-1–11301-13, 2012, with the permission of AIP Publishing

to manipulate fluids in a controlled way, using hydrophobic barriers to delimit the hydrophilic channels [26].

This patterning technique relies on the total hydrophobization of the substrate, followed by a selective dehydrophobization, where the hydrophilic channels form. The substrate is first soaked into the photoresist, spun to remove the excess photoresist, and pre-baked to remove the cyclopentanone in the SU-8 formula (epoxy-based photoresist using bisphenol-A diglycidyl ether resin) [38]. Then substrate and mask containing the pattern are aligned and irradiated with UV light to cure the photoresist. The irradiated substrate is then baked to cross-link the exposed regions of the photoresist. Soaking the substrate in a solvent (propylene glycol monomethyl ether acetate (PGMEA)) and washing with propan-2-ol remove the non-polymerized photoresist of the substrate (regions not irradiated by the UV light). Due to residues of photoresist in the hydrophilic regions, an extra step to increase the hydrophilicity of the paper is necessary (oxygen plasma treatment) [26]. The steps required to manufacture μPADs using lithography are represented in Fig. 4.9 [26].

The advantages of photolithography as a patterning method for μPAD manufacture include high channel and barrier resolution (186 ± 13 μm and 248 ± 13 μm, respectively [39]). However, the method disadvantages include the need for a mask for each microchip pattern, the high number of fabrication steps, and the high cost of equipment and reagents, hindering its use for fast prototyping and mass production.

Some strategies were developed to reduce costs associated with this fabrication method, including the change of the photoresist from SU-8 to SC (cyclized poly(isoprene) derivative) [38]. Besides its low cost (that can be lowered even further with photoresist dilution in xylenes), another advantage of this approach is the avoiding of pre- and post-baking steps [38]. However, due to a color change of the photoresist during curing, this method requires two subsequent exposures to UV light from both sides of the paper to fully cure the polymer and form the hydrophobic barriers [38]. That requires a process of alignment of two masks, which is undesirable for high-precision applications.

The Whitesides group proposed another strategy that is the preparation of a homemade version of SU-8 photoresist to diminish reagent costs, besides using adhesive transparency films instead of conventional photolithography masks to pattern the paper, in a method termed FLASH (Fast Lithographic Activation of Sheets) [39]. This method overcomes the limitations of the original lithographic process and enables its use for fast prototyping and mass production.

Other photolithographic methods appeared in the literature as variants of the original method [40–42]. The method that excels regarding channel and barrier resolution (~100 μm and ~60 μm, respectively) is the laser-based direct-writing technique on a nitrocellulose membrane soaked into an epoxy-based light-sensitive photopolymer, followed by a solvent wash to remove unpolymerized resin [43]. This method eliminates pre- and post-baking steps and plasma posttreatment, besides dispensing the use of a mask for patterning. However, it requires an x-y-z stage and a continuous-wave diode laser with the respective optical components, which increases production costs [40].

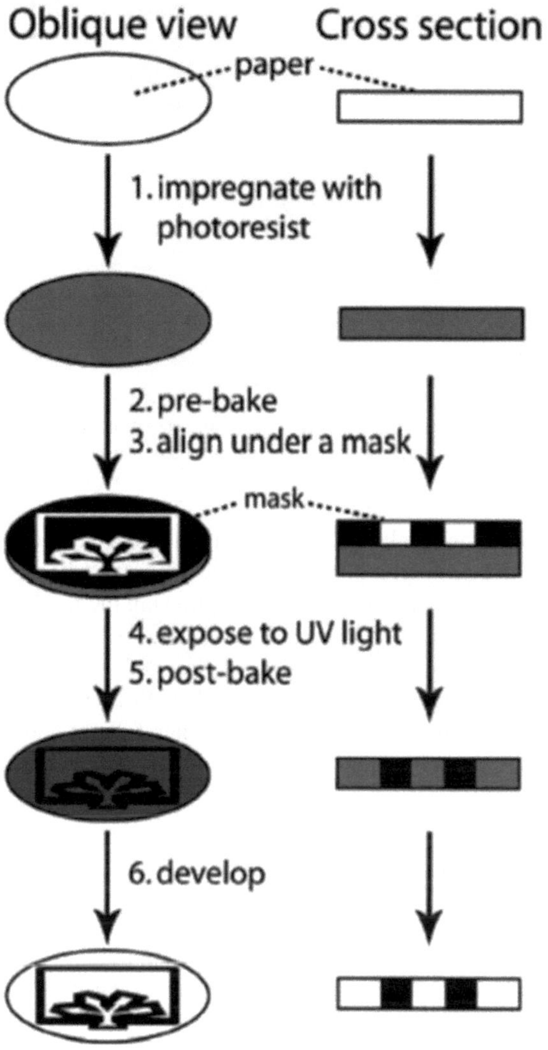

Fig. 4.9 Schematic representation of photolithography for μPAD manufacture. Reprinted with permission from ref. [7]. Copyright (2008) American Chemical Society

4.2.2 Plotting

In an attempt to minimize μPAD production costs, a one-step plotting method was introduced in 2012 by Nie et al. [28]. This patterning technique relies on the selective hydrophobization of the substrate, where the hydrophobic barriers are defined. Making use of an iron mask and a permanent marker pen, the authors have demonstrated the feasibility of fabricating hydrophobic barriers on paper, when the solvent of the permanent marker evaporates off paper delimiting then the hydrophobic channels. The manufacture scheme of μPADs using this method is represented in Fig. 4.10 [28].

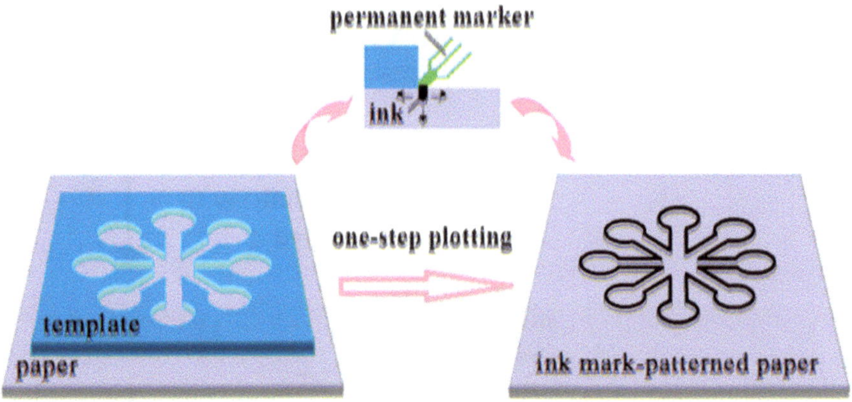

Fig. 4.10 Schematic representation of one-step plotting method for μPAD manufacture. Reprinted with permission from ref. [28]. Copyright (2012) American Chemical Society

Plotting is a low-cost and straightforward method for μPAD manufacture, ideal for fast prototyping. However, the process lacks automation because the barriers are drawn by hand with the pen following the mask pattern contours. Additionally, the reproducibility of the features (size of hydrophobic barriers and hydrophilic channels) and the resolution of the method are poor in comparison with other manufacture methods. μPADs produced with this approach are not compatible with common organic solvents [44], and therefore the application of these devices should be known beforehand.

4.2.3 Plasma Treatment

Plasma treatment to increase paper hydrophilicity for μPAD manufacture has been employed since the first paper-based microdevice work using photolithography [26]. However, plasma treatment as the primary step for μPAD production was first presented in 2008 by Li et al. [19]. This patterning method relies on the total hydrophobization of the substrate, followed by a selective dehydrophobization. The substrate is first immersed in an alkyl ketene dimer solution in *n*-heptane (AKD-heptane), letting the solvent evaporate in a fume hood. Then, baking the dry substrate at 100 °C for 45 min cures the AKD, making the substrate completely hydrophobic. The hydrophobic material is sandwiched between two metal masks cut with the hydrophilic channel pattern, and the whole system is introduced in a vacuum plasma chamber to oxidize the AKD selectively, forming the hydrophilic channels [19].

Compared to the conventional photolithographic method [26], plasma treatment has the main advantage with a smaller number of steps. However, this approach lacks versatility (one microdevice design per template), which limits its use in fast prototyping. μPADs fabricated with this method are not compatible with organic solvents like hexane, isooctane, and tetrahydrofuran (THF) [45].

4.2.4 *Wax Printing*

Wax printing for μPAD fabrication was introduced independently by Lu et al. [29] and Carrilho et al. [4] in 2009. Wax printing technology relies on the same principles as piezoelectric inkjet printers, with advantages including instant ink drying, no clogging of the printer head, and limited ink spreading after printing [46].

This technique creates a selective hydrophobization of paper by direct writing with hydrophobic ink, which is made of wax in this case. Wax patterns are printed on the surface of the paper, solidifying upon contact. Then, the sheet of paper containing the patterns is heated to melt the wax, which permeates the entire thickness of the paper creating the hydrophobic barriers and therefore the hydrophilic channels. The basic steps required to manufacture μPADs using wax printing are represented in Fig. 4.11 [4].

The advantages of using a wax printer for μPAD manufacture include high throughput (24 sheets per minute), low costs in comparison with other printing methods, and versatility of other useful substrates for printing [4, 47]. Wax printing has been the method of choice for μPAD fast prototyping [4, 48], and due to its intrinsic low cost and simplicity has potential to be used for mass production of μPADs.

The disadvantages of wax printing for μPAD production are the need for a thermal treatment [27] and the relatively poor resolution of the features in comparison with photolithography manufacturing methods [4]. For example, the dimensions for the final hydrophobic barriers and the final hydrophilic channels are 850 ± 50 μm and 561 ± 45 μm, respectively. In the past, the relatively high initial investment of a wax printer was one of the major disadvantages of the method [40], but now the price of this technology is minimal.

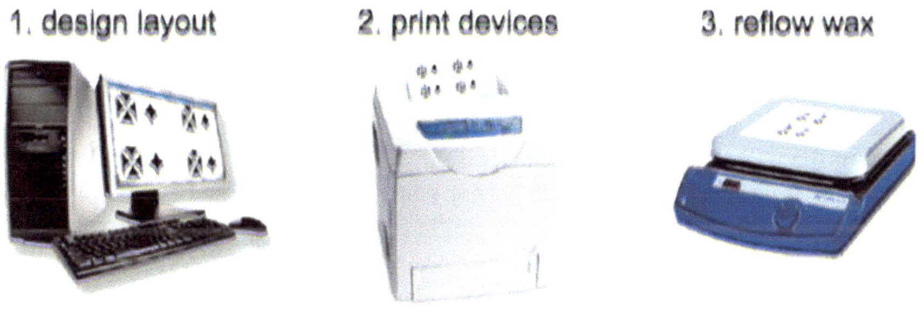

Fig. 4.11 Schematic representation of wax printing for μPAD manufacture. Reprinted with permission from ref. [4]. Copyright (2009) American Chemical Society

μPADs fabricated by wax printing are compatible with standard aqueous systems used with biological systems, including acidic solutions (HCl 0.1 mol L^{-1}), basic solutions (NaOH 0.1 mol L^{-1}), saline phosphate buffers, and glycerol. Such compatibility means that the hydrophobic barriers keep functional upon the addition of such solvents [4]. However, wax-printed μPADs are not compatible with organic solvents, including methanol, ethanol, *n*-propanol, methylene chloride, and xylenes [4].

4.2.5 Wax Dipping

Songjaroen et al. [30] proposed in 2011 another μPAD fabrication technique using wax in an attempt to lower the costs of wax printing method. As in wax printing, this technique consists of a selective hydrophobization of paper to generate hydrophobic barriers. The pattern mold is a ferromagnetic material cut in the shape of the hydrophilic channels, and placed on the substrate surface. The paper and the mold are placed on a glass slide, and a permanent magnet is placed behind the glass slide, holding the system together. The entire apparatus is then dipped into molten wax for 1 s, and the paper is allowed to cool to room temperature. The permanent magnet is removed, and the system is disassembled, leaving the hydrophilic channel on the paper otherwise covered with wax [30].

The biggest advantage of wax dipping for μPAD manufacture is its intrinsic low cost. However, the low hydrophilic channel resolution (639 ± 7 μm [30]) and the lack of versatility (one design per mold) limit the applicability of this method, being not suitable for either fast prototyping or mass production. Because the composition of the hydrophobic barriers is the same as in wax printing method, this method also presents the same solvent limitations, being incompatible with common organic solvents [4].

4.2.6 Inkjet Printing

Inkjet printing technology, proposed by Maejima et al. [31] in 2013, appeared as an innovative method for μPAD manufacture. This method delivers a selective hydrophobization of paper using only an inkjet printer and a UV light source. Most of the μPAD fabrication methods do not allow for the manufacture of a fully functional microchip (hydrophobic barrier patterning and reagent deposition steps) using a single piece of equipment or a single process. Thus, inkjet printing rises as a tool to make it possible.

The ink dispensed by the printer is composed of a hydrophobic UV-curable acrylate (monomers of octadecyl acrylate and 1,10-decanediol diacrylate) made of nonvolatile compounds. The cartridge designed for the black ink is then filled with this composition. Both sides of the substrate (topside and backside) are printed with

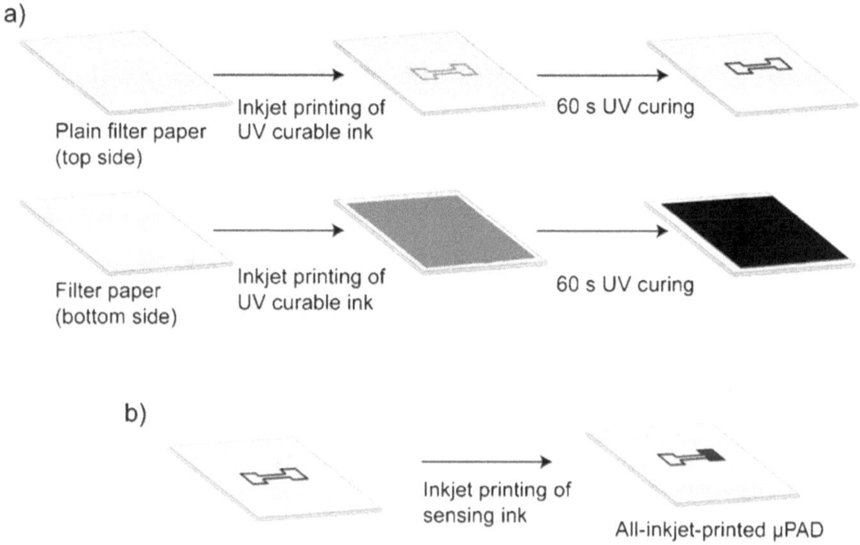

Fig. 4.12 Schematic representation of inkjet printing for μPAD manufacture process. (**a**) Patterning of the substrate by inkjet printing in both sides (gray color indicates the printed hydrophobic features before UV curing, and black color indicates the barrier after curing). (**b**) Inkjet printing of the reagents for the assays. Reproduced from ref. [31] with permission of The Royal Society of Chemistry

the ink and are exposed to a UV light source to cure the polymer during 60 s. The cartridge designed for yellow ink is filled with a solution of reagents specific for the assay being conducted, which is printed in the testing areas of the microdevices [31]. The steps required to manufacture μPADs using inkjet printing are represented in Fig. 4.12.

The advantages of this method include its efficiency for the production of fully functional microchips with a single equipment and high microfluidic channel resolution (272 ± 19 μm) [31]. Inkjet printing presents great potential to be used for mass production of μPADs, and the initial low-cost investment of equipment allows for μPAD fast prototyping as well. The disadvantages of this method include the toxicity of reagents used in the ink and the use of organic solvents, similar to photolithographic methods [31, 49].

4.2.6.1 Inkjet Etching

An indirect patterning method to fabricate μPADs using an inkjet printer was developed by Abe et al. [32] in 2008, also known as inkjet etching. This patterning technique produces a total hydrophobization of the substrate, followed by a selective dehydrophobization to create the hydrophilic channels. This microfabrication method involves soaking the substrate with a solution of poly(styrene) dissolved

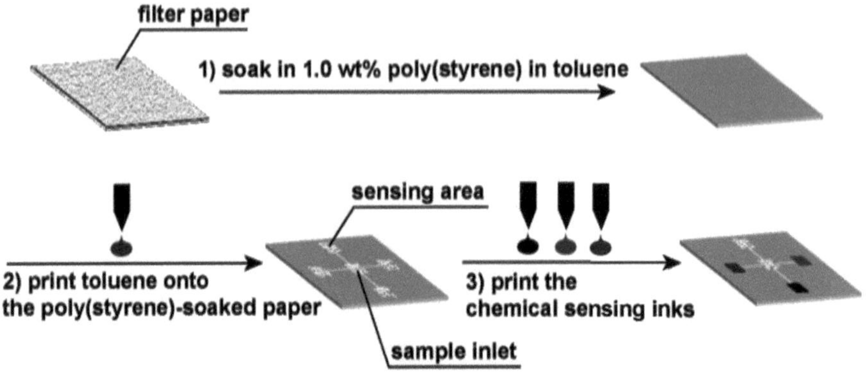

Fig. 4.13 Schematic representation of inkjet etching for μPAD fabrication. The cellulosic substrate is modified to be hydrophobic (step 1). Steps 2 (patterning) and 3 (chemical-sensing reagent printing) are performed using the same inkjet printing apparatus. Reprinted with permission from ref. [32]. Copyright (2008) American Chemical Society

in toluene during 2 h, and then drying it at room temperature during 15 min, which creates the hydrophobic substrate. Printing the desired layout with toluene on the hydrophobic substrate using an adapted inkjet printer creates the hydrophilic channels. This enables dissolution of the polymeric material and therefore the exposition of the hydrophilic paper, forming the hydrophilic channels [49]. Figure 4.13 summarizes the main three steps required to manufacture μPADs using inkjet etching.

The advantages of this technology include dispensing of sophisticated additional equipment, the use of expensive photoresist reagents, and the fabrication of a fully functional microchip using a single modified inkjet printer, like in the inkjet printing fabrication method.

The disadvantages of inkjet etching are the use of organic solvents such as toluene, which is a flammable and volatile organic compound (VOC) [49], besides the requirement of ten print cycles for etching the hydrophilic patterns, increasing the manufacture time of the devices. The hydrophilic channels generated with this method present relatively poor resolution (~550 μm) in comparison with photolithography manufacture method. This method is more appropriate for fast prototyping of μPADs, and due to its disadvantages this is an unfeasible method for large-scale production of microdevices.

4.2.7 Flexography Printing

Most techniques to produce paper-based microdevices are focused on benchtop and at-a-site applications, most probably because this new technology is still in the development of concept stage, and to achieve an industrial level production a

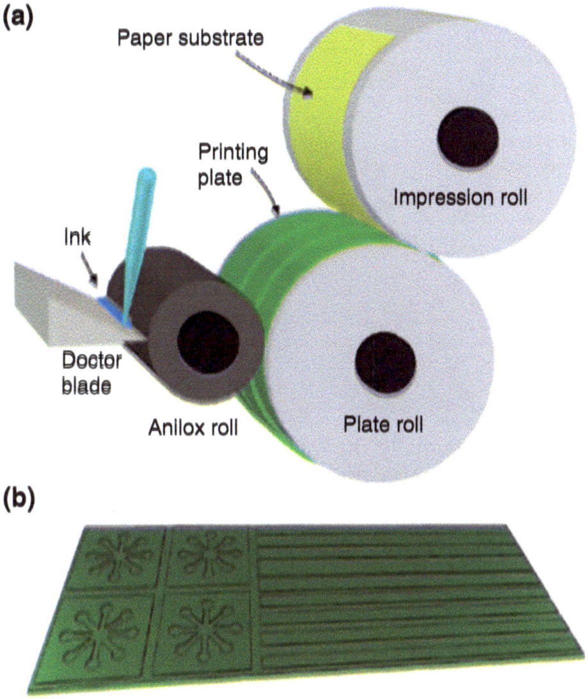

Fig. 4.14 Schematic representation of the flexography printing for μPAD fabrication. (**a**) Flexography unit to manufacture μPADs. (**b**) Relief patterns that define hydrophobic barriers on the microchip. Reprinted with permission from ref. [33]. Copyright (2010) American Chemical Society

series of optimizations and scaling processes are still required. Olkkonen et al. [33] in 2010 described the flexographically printed technique for μPAD manufacture, envisioning a large-scale production of such devices.

Flexography printing is based on the selective hydrophobization of the substrate. Hydrophobic barriers are created by applying a polystyrene solution on an anilox ceramic roll, and the hydrophobic ink is transferred to a relief pattern attached to the printing plate. The paper substrate is then put into contact with the printing plate, and the ink is transferred to the paper, forming the hydrophobic barriers and the hydrophilic channels [33]. The steps required to manufacture μPADs using flexography printing are detailed in Fig. 4.14.

The advantages of this method to produce μPADs are its high throughput (60 m of substrate per minute) and the low number of steps required, ideal for mass production of devices [33]. Its disadvantages include a lower resolution of hydrophilic channels (500 ± 30 μm) than photolithography, and the need for different relief patterns for different device designs, which hinders its use for fast prototyping at the development stage.

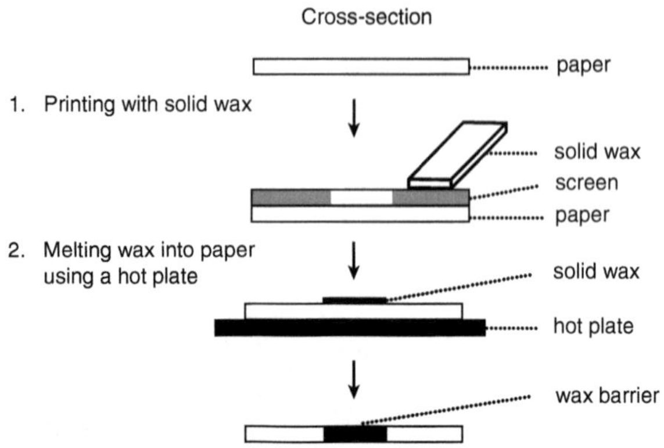

Fig. 4.15 Schematic representation of the fabrication of μPADs using wax screen printing process. Reproduced from ref. [34] with permission of The Royal Society of Chemistry

This technology is potentially the one to be implemented in the industry with minor adjustments that could facilitate the manufacture process, which includes automatic deposition of ink, an automated system for ink transfer roller washing, as well as a collection mechanism for the waste generated during the process.

4.2.8 Screen Printing

The use of screen printing for μPAD fabrication was first presented in 2011 by Dungchai et al. [34], based on the selective hydrophobization of the substrate to create the hydrophobic barriers. It was an attempt to minimize μPAD production costs by wax printing technology [4], mainly the initial investment to purchase the printer, but also economizing in commercial wax sticks, equipment maintenance, and energy consumption.

In the wax screen printing technique, the preparation of the screen consists of designing and cutting a pattern on a transparency film, which makes contact with the paper substrate. Rubbing a solid wax stick against the assemble deposits the wax on the paper surface with the pattern cut in the screen [34]. The wax melts when the paper is on a hot plate, creating the desired hydrophobic barriers. The manufacturing process by wax screen printing is schematized in Fig. 4.15 [34].

The advantages of this patterning method include its intrinsic low cost, dispensing the initial investment of a wax printer, and the ability to cut quickly different patterns into transparency sheets, making it versatile for simple-design microchip fast prototyping. However, its disadvantages include low-resolution hydrophilic channels and hydrophobic barriers (650 μm and 1300 μm, respectively), besides an extra thermal treatment step [34]. Although this method is relatively inexpensive, for large-scale production it would require major optimizations, similar to processes of silkscreen printing on clothing. This method presents the same solvent limitations of wax printing, being not compatible with common organic solvents [4].

Another strategy using screen printing technology uses a solution of polystyrene in toluene to create the hydrophobic barriers instead of wax and a solvent-compatible polyester screen [50]. This method does not require a thermal treatment step after applying the hydrophobic material on the substrate; it requires just a resting period for solvent evaporation. This method provides higher resolution features than the wax screen printing technology (hydrophilic channel = 671 ± 50 μm and hydrophobic barrier = 380 ± 40 μm), being an improvement of the previous technology.

4.2.9 *Laser Treatment*

Laser treatment for μPAD fabrication was introduced in 2011 by Chitnis et al. [35], based on the selective dehydrophobization of the paper substrate to form hydrophilic channels. The initial substrates used in this work consisted of commercially available surface-treated hydrophobic papers, such as parchment paper, making this a one-step fabrication method. A CO_2 laser cutting/engraving machine is used to treat the surface of the paper substrate, and the heat produced by the laser changes the wettability of the surface of the hydrophobic papers [35].

The main advantages of this method rely on its high throughput (laser raster scan speeds of 4 ft^2 per hour, depending on the equipment), the use of commercially available hydrophobic substrates, and the high resolution of the features (hydrophilic channel = 62 ± 1 μm and hydrophobic barrier = 250 μm). Also, because such devices start with a polymeric coated paper substrate, there is a gain in mechanical strength in the μPAD structure [35]. Because the hydrophobic agent is already present in the paper, there is no need for a solvent washing, being an improvement of the photolithographic laser-based direct-write technique [35].

It is relevant to notice that microchannels created using this method do not allow for the transport of fluids through them. For applications in which just the containment of reaction zones is needed, such as paper-based enzyme-linked immunosorbent assay (p-ELISA) microplates [51, 52], this method is ideal for large-scale production of devices. However, if the transport of fluids in the matrix is required, then a silica nanoparticle coating is needed, which is a manual process hindering its use as a large-scale manufacture method.

4.2.10 *Stamping*

Stamping for μPADs fabrication was introduced in 2013 by Curto et al. [36]. This technique is based on the selective creation of hydrophobic barriers on the substrate [36, 37, 40, 53]. Commercial permanent ink (Black Lumocolor®) is used as a modifying agent, being applied to a polydimethylsiloxane (PDMS)-based stamp by gently pushing it against the ink-saturated stone pad. The indelible ink is transferred from the PDMS stamp to the paper substrate by contact, creating microfluidic devices in less than 10 s [36], being the greatest advantage of this method. This

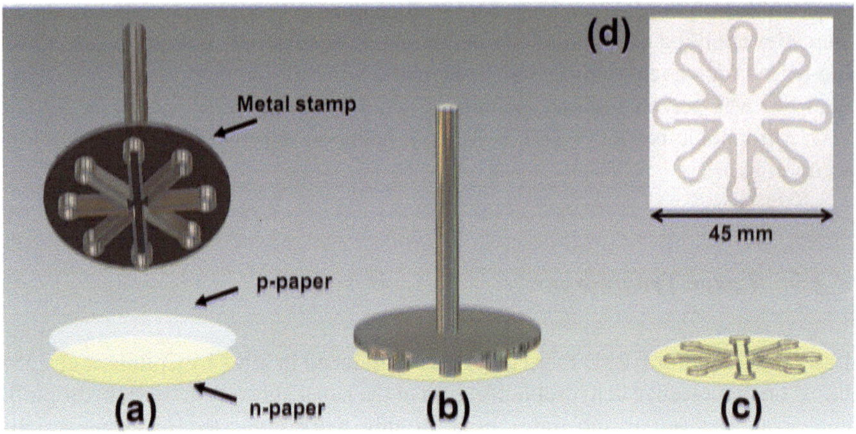

Fig. 4.16 Fabrication process of μPADs using the wax stamping technique. (**a**) A paraffinized paper substrate is placed over the native paper surface. (**b**) A metal stamp containing the desired pattern is heated at 150 °C and brought into contact with the stacked papers. (**c**) Depiction of a device fabricated by this method. (**d**) A μPAD created with this method. Reproduced from ref. [37] with permission of The Royal Society of Chemistry

technique has great potential to be applied to manufacture of μPADs in large scale. However, the low-resolution features (hydrophobic barriers >1100 μm) and the need for different stamps for different designs limit its application as an exploratory and fast prototyping tool [36].

Another strategy for μPAD fabrication based on stamping was proposed in 2014 by Garcia et al. [37], also based on the selective hydrophobization of the substrate. The stamp containing the desired pattern is machined on stainless steel. Paraffinized paper (p-paper) is prepared by soaking a sheet of paper in melted paraffin and allowing it to dry. The p-paper is placed over a normal paper substrate (n-paper) and the metal stamp is heated to 150 °C and placed on the stacked system, causing the paraffin of p-paper to permeate into the n-paper, creating the microchannels in the latter. This process is represented in Fig. 4.16 [37].

The biggest advantage of this method is its low cost, which could benefit small-scale production in regions with lack of resources. However, the lower resolution of the features generated by this method (hydrophilic channel = 2.6 ± 0.1 mm and hydrophobic barrier = 1.4 ± 0.1 mm) hinders its application for large-scale production that requires elaborate designs. Also, the need for a different metal stamp for each design hinders the use in exploratory studies [37]. Because this method is based on wax to form the hydrophobic barriers, it presents the same solvent limitations of wax printing technology [4].

4.2.11 Other Methods

The simplest approach to fabricate channels for fluidic manipulation in a cellulosic matrix consists of cutting the patterns onto the substrate [2, 54], but this method presents as drawbacks with channel size [40] and microchips with simple designs, besides making it difficult to handle the device, which can hinder many applications. However, this method is encouraged when there is incompatibility of solvents with the hydrophobic barriers in the other methods, being ideal for the analysis of samples in organic solvents.

An innovative method for μPAD fabrication using corona discharge treatment has been proposed [55]. The handheld corona treatment uses a paper substrate that is previously hydrophobized using a solution of octadecyltrichlorosilane (OTS). Then a mask containing the desired pattern is stacked on top of the OTS paper, and the system is exposed to a corona discharge, creating the microchannels [55]. This manufacture method is a promising technology to create μPADs, but is still early in its development.

Another innovative approach includes the selective hydrophobization of the substrate using the deposition of polymers in the vapor phase [56]. The monomer is placed in a sublimator chamber, being converted by pyrolysis in a radical monomer, which is deposited and polymerized on paper's surface. A mask is placed on paper's surface with the design of the device, allowing for the selective modification of the substrate [56]. This technology employs sophisticated equipment, which impacts the final cost of the device, making it difficult to be implemented in an industrial scale.

4.3 Economics: What to Consider

Before choosing a fabrication method to manufacture μPADs, it is necessary to know beforehand important features and specifications. For example, it is important to define the intended application for a given device and the characteristics of the sample being analyzed (solvent composition), the design of the device, and the feature requirements (resolution of the hydrophilic channels). In addition, before decision, it should be taken into consideration if the fabrication method permits easily exploratory studies (fast prototyping), and if the final application requires benchtop or industrial scale. The choice of the manufacture method also depends on factors such as (1) the initial investment required, which includes the purchase of the equipment (wax printer, inkjet printer, CO_2 laser cutter/engraving machine, UV curing chamber) or the fabrication/adaptation of equipment (metal molds for stamps, inkjet printer modification, photolithographic masks); (2) reagent consumption and toxicity; (3) energy consumption to fabricate the devices; and (4) the operator costs. Table 4.2 summarizes the main features of a number of ways of

Table 4.2 Main substrates used for the creation of hydrophobic barriers on paper

Material	Fabrication method	Process	Advantages	Disadvantages	Relative cost	Ref.
Wax/paraffin	Wax printing, screen printing, wax dipping, and stamping	Wax permeates through the entire thickness of paper after melting	Dispenses the use of solvent or need of curing steps; direct processing without leaving residues on channels (test zone)	Requires an extra thermal processing	Low (paraffin) High (wax sticks for wax printer)	[4, 37]
Polystyrene	Etching, screen printing, flexography, and inkjet printing	Polystyrene forms a liquid-guiding barrier after solvent evaporation	Potential for large-scale production and versatility	Use of organic solvents and generation of toxic waste; leaves residues of polymers on channels	Low	[33]
Poly(dimethyl-siloxane) (PDMS)	Plotting and flexography	PDMS in hexane wets and permeates through the entire thickness of the paper	Flexibility due to elastomeric characteristics	Needs an extra curing step; generation of waste	Low	[57]
SU-8 photoresist	Photolitho-graphy	Paper is soaked into the photoresist, followed by a pre-baking process, and is irradiated with UV light to cure the photoresist	A well-known technique that provides high resolution to channels	Use of organic solvents and generation of toxic waste; requires extra washing and baking steps, besides plasma treatment	High	[38]
Octadecyltrichloro silane(OTS)	Etching and handheld corona treatment	Paper is soaked into the OTS solution in *n*-hexane, and OTS molecules are coupled to the cellulose fibers	Enables the use of devices in both aqueous and organic solvents	Leaves polymer residues on channels	Low	[55, 58]

Hydrophobic sol-gel	Etching and inkjet printing	Paper is soaked in a solution of polymers that form a hydrophobic cross-linked siloxane network	Resistant to surfactant solutions	Generation of waste; leaves residues of polymers on channels	Low	[59]
Silicone	Inkjet printing	The hydrophobic silicone resin barriers on the paper are formed using a silicone precursor (siloxanes)	Flexibility, which is due to elastomeric characteristics and resistant to a variety of aqueous surfactant solutions and to some organic solvents	Requires the use of organic solvents; generation of waste	Low	[60]
Polyacrylate	Inkjet printing	Hydrophobic acrylate monomers of octadecyl acrylate and 1,10-decanediol diacrylate are deposited on the paper, penetrating through its entire thickness and are cured by exposition to a UV light source	Nonvolatile and resistant against liquids with lowered surface tension	Needs an extra curing step; generation of waste	High	[31]

producing the hydrophobic barriers on the porous substrates, presenting pros and cons, and the relative cost of each one.

The paper substrate itself plays a significant role in the costs associated with the μPAD manufacture. The preferred choice in most paper-based device studies is chromatography paper from Whatman®—1 CHR—which enables better reproducibility due to its uniformity and composition [4, 8, 10, 16, 20, 25, 39, 61]. It is ideal for the starting point of exploratory studies and for benchtop production scale where paper consumption is low. However, if the demand calls for large-scale production, then other, lower cost types of paper should be considered, taking into account that each paper has specific physical-chemical properties that can affect assay results [62]. In the choice of the paper substrates, one should consider aspects as pore size, thickness, weight, and wettability. For example, the porosity and thickness of paper affect sample permeation through the device, which can be problematic for more viscous samples [17]. The weight and thickness of the paper substrate should also be considered when choosing the manufacturing method. Such features are important because some papers may not be compatible with methods like wax printing and inkjet printing.

Although paper substrates are known to present high wettability (chromatography and filter papers), some variations of cellulosic substrates can present different properties. For example, nitrocellulose membrane has very small pore sizes and cellulose acetate can be blended in its matrix, making a material with hydrophobic characteristics that is widely used in commercial lateral flow tests [1, 63–65]. Other papers with low wettability such as office paper [66] and glossy paper [67] are also being used as μPAD substrates in specific applications, like electrochemical sensors [66]. Table 4.3 summarizes the main types of paper substrates used in μPAD manufacture.

4.4 Scalability and Cost

Although all methods described in Sect. 4.2 are capable of manufacturing paper-based devices, not all of them are suitable for mass production. The scalability of a μPAD manufacture process depends on all the costs associated with the process (Sect. 4.3), the throughput of the process in terms of speed of device fabrication, and the required resolution of the features of devices, all of which will impact the final price of the device.

Throughout previous sections, we have detailed the methods of μPAD fabrication that have potential or are already suitable for large-scale production, oppositely to other methods that are still limited to benchtop application. Table 4.4 summarizes all these methods. Currently, the process that has the greatest potential to be used for mass production of paper-based devices (only considering the paper matrix to act as a support to contain and manipulate fluids) is the flexography printing. Flexography requires just a few minor adjustments from current technology and presents the highest throughput of the existing methods (one-step processing on paper with 60 m

Table 4.3 Features and characteristics of main paper substrates used in μPAD manufacturing

Paper substrate		Properties	Relative cost[a]	Ref.
Chromatographic paper	Whatman® Grade 1 CHR	Thickness (μm): 180 Pore size/particle retention (μm): 11 Weight (g/m²): 87 Classification: flow rate of 130 mm/30 min	High	[8, 38]
	Whatman® Grade 3MM CHR	Thickness: 340 Pore size/particle retention: 6 Weight (g/m²): 189 Classification: flow rate of 130 mm/30 min	High	[8]
	Whatman® Grade 4 CHR	Thickness: 210 Pore size/particle retention: 20 Weight (g/m²): 93 Classification: flow rate of 180 mm/30 min	High	[8]
Filter paper	Whatman® Grade 1	Thickness (μm): 180 Pore size/particle retention (μm): 11 Weight (g/m²): 87 Classification: medium flow	Low	[8]
	Whatman® Grade 3	Thickness (μm): 390 Pore size/particle retention (μm): 6 Weight (g/m²): 185 Classification: medium flow	Low	[8]
	Whatman® Grade 4	Thickness (μm): 205 Pore size/particle retention (μm): 20–25 Weight (g/m²): 92 Classification: very fast flow	Low	[8, 68]
	Whatman® Grade 5	Thickness (μm): 100 Pore size/particle retention (μm): 2.5 Weight (g/m²): 200 Classification: slow flow	Low	[68]
Nitrocellulose		Pore size (μm): 0.45 Can present hydrophobic characteristics if blended with cellulose acetate	High	[1, 62, 63]
Glossy paper		Low wettability/high hydrophobicity	Low	[67]
Office paper		Low wettability/high hydrophobicity	Low	[66]

[a]The cost of each material varies largely upon quantity

Table 4.4 Current most common μPAD manufacturing methods and their versatility towards scale of production and product development

μPADs fabrication techniques	Relative cost	Fast prototyping	Mass production	Ref.
Photolithography	High	✓[a]	✓	[26, 39]
Plotting	Low	✓		[28]
Plasma treatment	High			[19]
Wax printing	Low	✓	✓	[4, 29]
Wax dipping	Low			[30]
Inkjet printing	Low	✓	✓	[31]
Inkjet etching	Low	✓		[32]
Flexography printing	High		✓	[33]
Screen printing	Low	✓	✓	[34]
Laser treatment	High	✓	✓[b]	[35]
Stamping	Low		✓	[36, 37]

[a]It depends on the photoresist used
[b]This method is just recommended for mass production of devices that do not require on-chip fluidic manipulation, i.e., paper-based ELISA plates

of the substrate processed per minute) [33]. However, for a fully functional device, then the process indicated for mass production of μPADs is inkjet printing. Inkjet printing uses the same equipment to pattern the paper substrate and to deposit the reagents for the assay in colorimetric reactions [31]. Additionally, inkjet printers can also print electrodes on paper [69] for electrochemical detection [66].

4.5 Summary

This chapter aimed to cover all the key points for the manufacture of paper-based microfluidic analytical devices (μPADs). We introduced the importance of the design of the microdevices and the economic aspects to consider when fabricating them. Moreover, throughout this chapter, we discussed the current primary technologies and alternative methods to manufacture μPADs, enabling the reader to understand how to fabricate these microdevices, and which method to select for different applications. We discussed many aspects including (1) initial investment, (2) choice of papers, and (3) scalability throughout this chapter. Thus, all these perspectives will enable the reader to have a holistic view concerning the production of μPADs, helping researchers and technicians who work with this technology to select the best fit for their needs.

Acknowledgements The authors would like to thank the funding agencies FAPESP (Grant No. 2011/13997-8), CNPq (Grant No. 205453/2014-7) by the scholarships and the financial support to the Instituto Nacional de Ciência e Tecnologia de Bioanalítica—INCTBio (FAPESP Grant Nr. 2008/57805-2/CNPq Grant Nr. 573672/2008-3), the Georgia Institute of Technology (Georgia Tech), and the State of Georgia, USA. The authors declare having no competing financial interests.

References

1. Yetisen AK, Akram MS, Lowe CR (2013) Paper-based microfluidic point-of-care diagnostic devices. Lab Chip 13:2210–2251
2. Cate DM, Adkins JA, Mettakoonpitak J, Henry CS (2015) Recent developments in paper-based microfluidic devices. Anal Chem 87:19–41
3. Xia Y, Si J, Li Z (2016) Fabrication techniques for microfluidic paper-based analytical devices and their applications for biological testing: a review. Biosens Bioelectron 77:774–789
4. Carrilho E, Martinez AW, Whitesides GM (2009) Understanding wax printing: a simple micropatterning process for paper-based microfluidics. Anal Chem 81:7091–7095
5. Morbioli GG, Mazzu-Nascimento T, Milan LA, Stockton AM, Carrilho E (2017) Improving sample distribution homogeneity in three-dimensional microfluidic paper-based analytical devices by rational device design. Anal Chem 89:4786–4792
6. Peeling RW, Holmes KK, Mabey D, Ronald A (2006) Rapid tests for sexually transmitted infections (STIs): the way forward. Sex Transm Infect 82(Suppl 5):v1–v6
7. Martinez AW, Phillips ST, Carrilho E, Thomas SW, Sindi H, Whitesides GM (2008) Simple telemedicine for developing regions: camera phones and paper-based microfluidic devices for real-time, off-site diagnosis. Anal Chem 80:3699–3707
8. Evans E, Gabriel EFM, Coltro WKT, Garcia CD (2014) Rational selection of substrates to improve color intensity and uniformity on microfluidic paper-based analytical devices. Analyst 139:2127–2132
9. Elizalde E, Urteaga R, Berli CLA (2015) Rational design of capillary-driven flows for paper-based microfluidics. Lab Chip 15:2173–2180
10. Han SI, Hwang KS, Kwak R, Lee J (2016) Microfluidic paper-based biomolecule preconcentrator based on ion concentration polarization. Lab Chip 16:2219
11. Wong SY, Cabodi M, Rolland J, Klapperich CM (2014) Evaporative concentration on a paper-based device to concentrate analytes in a biological fluid. Anal Chem 86:11981–11985
12. Martinez AW, Phillips ST, Whitesides GM (2008) Three-dimensional microfluidic devices fabricated in layered paper and tape. Proc Natl Acad Sci U S A 105:19606–19611
13. Osborn JL, Lutz B, Fu E, Kauffman P, Stevens DY, Yager P (2010) Microfluidics without pumps: reinventing the T-sensor and H-filter in paper networks. Lab Chip 10:2659
14. Oh KW, Lee K, Ahn B, Furlani EP (2012) Design of pressure-driven microfluidic networks using electric circuit analogy. Lab Chip 12:515–545
15. Lewis GG, DiTucci MJ, Baker MS, Phillips ST (2012) High throughput method for prototyping three-dimensional, paper-based microfluidic devices. Lab Chip 12:2630–2633
16. Zhu WJ, Feng DQ, Chen M, Chen ZD, Zhu R, Fang HL, Wang W (2014) Bienzyme colorimetric detection of glucose with self-calibration based on tree-shaped paper strip. Sensors Actuators B Chem 190:414–418
17. Lutz B, Liang T, Fu E, Ramachandran S, Kauffman P, Yager P (2013) Dissolvable fluidic time delays for programming multi-step assays in instrument-free paper diagnostics. Lab Chip 13:2840–2847
18. Houghtaling J, Liang T, Thiessen G, Fu E (2013) Dissolvable bridges for manipulating fluid volumes in paper networks. Anal Chem 85:11201–11204
19. Li X, Tian J, Nguyen T, Shen W (2008) Paper-based microfluidic devices by plasma treatment. Anal Chem 80:9131–9134
20. Schilling KM, Lepore AL, Kurian JA, Martinez AW (2012) Fully enclosed microfluidic paper-based analytical devices. Anal Chem 84:1579–1585
21. Morbioli GG, Mazzu-Nascimento T, Stockton AM, Carrilho E (2017) Technical aspects and challenges of colorimetric detection with microfluidic paper-based analytical devices (μPADs)—a review. Anal Chim Acta 970:1–22
22. Bhakta SA, Borba R, Taba M, Garcia CD, Carrilho E (2014) Determination of nitrite in saliva using microfluidic paper-based analytical devices. Anal Chim Acta 809:117–122

23. Liu H, Crooks RM (2011) Three-dimensional paper microfluidic devices assembled using the principles of origami. J Am Chem Soc 133:17564–17566
24. Connelly JT, Rolland JP, Whitesides GM (2015) Paper machine for molecular diagnostics. Anal Chem 87:7595–7601
25. Schilling KM, Jauregui D, Martinez AW (2013) Paper and toner three-dimensional fluidic devices: programming fluid flow to improve point-of-care diagnostics. Lab Chip 13:628–631
26. Martinez AW, Phillips ST, Butte MJ, Whitesides GM (2007) Patterned paper as a platform for inexpensive, low-volume, portable bioassays. Angew Chem Int Ed 46:1318–1320
27. Li X, Ballerini DR, Shen W (2012) A perspective on paper-based microfluidics: current status and future trends. Biomicrofluidics 6:11301–1130113
28. Nie J, Zhang Y, Lin L, Zhou C, Li S, Zhang L, Li J (2012) Low-cost fabrication of paper-based microfluidic devices by one-step plotting. Anal Chem 84:6331–6335
29. Lu R, Shi W, Jiang L, Qin J, Lin B (2009) Rapid prototyping of paper-based microfluidics with wax for low-cost, portable bioassay. Electrophoresis 30:1497–1500
30. Songjaroen T, Dungchai W, Chailapakul O, Laiwattanapaisal W (2011) Novel, simple and low-cost alternative method for fabrication of paper-based microfluidics by wax dipping. Talanta 85:2587–2593
31. Maejima K, Tomikawa S, Suzuki K, Citterio D (2013) Inkjet printing: an integrated and green chemical approach to microfluidic paper-based analytical devices. RSC Adv 3:9258
32. Abe K, Suzuki K, Citterio D (2008) Inkjet-printed microfluidic multianalyte chemical sensing paper. Anal Chem 80:6928–6934
33. Olkkonen J, Lehtinen K, Erho T (2010) Flexographically printed fluidic structures in paper. Anal Chem 82:10246–10250
34. Dungchai W, Chailapakul O, Henry CS (2011) A low-cost, simple, and rapid fabrication method for paper-based microfluidics using wax screen-printing. Analyst 136:77–82
35. Chitnis G, Ding Z, Chang C-L, Savran CA, Ziaie B (2011) Laser-treated hydrophobic paper: an inexpensive microfluidic platform. Lab Chip 11:1161–1165
36. Curto VF, Lopez-Ruiz N, Capitan-Vallvey LF, Palma AJ, Benito-Lopez F, Diamond D (2013) Fast prototyping of paper-based microfluidic devices by contact stamping using indelible ink. RSC Adv 3:18811
37. Garcia PDT, Cardoso TMG, Garcia CD, Carrilho E, Coltro WKT (2014) A handheld stamping process to fabricate microfluidic paper-based analytical devices with chemically modified surface for clinical assays. RSC Adv 4:37637–37644
38. Carrilho E, Phillips ST, Vella SJ, Martinez AW, Whitesides GM (2009) Paper microzone plates. Anal Chem 81:5990–5998
39. Martinez AW, Phillips ST, Wiley BJ, Gupta M, Whitesides GM (2008) FLASH: a rapid method for prototyping paper-based microfluidic devices. Lab Chip 8:2146–2150
40. He Y, Wu Y, Fu JZ, Wu WB (2015) Fabrication of paper-based microfluidic analysis devices: a review. RSC Adv 5:78109–78127
41. OuYang L, Wang C, Du F, Zheng T, Liang H (2014) Electrochromatographic separations of multi-component metal complexes on a microfluidic paper-based device with a simplified photolithography. RSC Adv 4:1093
42. He Y, Wu W, Fu J (2015) Rapid fabrication of paper-based microfluidic analytical devices with desktop stereolithography 3D printer. RSC Adv 5:2694–2701
43. He PJW, Katis IN, Eason RW, Sones CL (2015) Laser-based patterning for fluidic devices in nitrocellulose. Biomicrofluidics 9
44. Wang X (2008) US Patent 8,741,039. https://doi.org/10.1128/AEM.67.1.82-88.2001
45. Kumar S, Chauhan VS, Chakrabarti SK (2012) Separation and analysis techniques for bound and unbound alkyl ketene dimer (AKD) in paper: a review. Arab J Chem 9:S1636–S1642
46. Jaeger CW (2000) How does a solid ink printer work? http://www.imaging.org/ist/resources/tutorials/solid_ink.cfm. Accessed 3 Oct 2015
47. Jacob JAM (2013) Desenvolvimento de placas de microtitulação em papel. Universidade Nova de Lisboa, Lisboa

48. Morbioli GG (2015) Funcionalização de celulose para ensaios bioanalíticos em dispositivos microfluídicos baseados em papel (μPADs). Universidade de São Paulo, São Paulo
49. Yamada K, Henares TG, Suzuki K, Citterio D (2015) Paper-based inkjet-printed microfluidic analytical devices. Angew Chem Int Ed 54:5294–5310
50. Sameenoi Y, Nongkai PN, Nouanthavong S, Henry CS, Nacapricha D (2014) One-step polymer screen-printing for microfluidic paper-based analytical device (μPAD) fabrication. Analyst 139:6580–6588
51. Mazzu-Nascimento T, Morbioli GG, Milan LA, Donofrio FC, Mestriner CA, Carrilho E (2017) Development and statistical assessment of a paper-based immunoassay for detection of tumor markers. Anal Chim Acta 950:156–161
52. Mazzu-Nascimento T, Morbioli GG, Milan LA, Silva DF, Donofrio FC, Mestriner CA, Carrilho E (2017) Improved assessment of accuracy and performance indicators in paper-based ELISA. Anal Methods 9:2644–2653
53. Zhang Y, Zhou C, Nie J, Le S, Qin Q, Liu F, Li Y, Li J (2014) Equipment-free quantitative measurement for micro fluidic paper-based analytical devices fabricated using the principles of movable-type printing. Anal Chem 86:2005–2012
54. Wang W, Wu W-Y, Zhu J-J (2010) Tree-shaped paper strip for semiquantitative colorimetric detection of protein with self-calibration. J Chromatogr A 1217:3896–3899
55. Jiang Y, Hao Z, He Q, Chen H (2016) A simple method for fabrication of microfluidic paper-based analytical devices and on-device fluid control with a portable corona generator. RSC Adv 6:2888–2894
56. Demirel G, Babur E (2014) Vapor-phase deposition of polymers as a simple and versatile technique to generate paper-based microfluidic platforms for bioassay applications. Analyst 139:2326–2331
57. Bruzewicz DA, Reches M, Whitesides GM (2008) Low-cost printing of poly(dimethylsiloxane) barriers to define microchannels in paper. Anal Chem 80:3387–3392
58. He Q, Ma C, Hu X, Chen H (2013) Method for fabrication of paper-based microfluidic devices by alkylsilane self-assembling and UV/O_3-patterning. Anal Chem 85:1327–1331
59. Wang J, Monton RN, Zhang X, Filipe CDM, Pelton R, Monton MRN, Zhang X, Filipe CDM, Pelton R, Brennan JD (2014) Hydrophobic sol-gel channel patterning strategies for paper-based microfluidics. Lab Chip 14:691–695
60. Rajendra V, Sicard C, Brennan JD, Brook MA (2014) Printing silicone-based hydrophobic barriers on paper for microfluidic assays using low-cost ink jet printers. Analyst 139:6361–6365
61. Li X, Tian J, Garnier G, Shen W (2010) Fabrication of paper-based microfluidic sensors by printing. Colloids Surf B Biointerfaces 76:564–570
62. Liana DD, Raguse B, Justin Gooding J, Chow E (2012) Recent advances in paper-based sensors. Sensors 12:11505–11526
63. Fridley GE, Holstein CA, Oza SB, Yager P (2013) The evolution of nitrocellulose as a material for bioassays. MRS Bull 38:326–330
64. Lu Y, Shi W, Qin J, Lin B (2010) Fabrication and characterization of paper-based microfluidics prepared in nitrocellulose membrane by wax printing. Anal Chem 82:329–335
65. Byrnes S, Thiessen G, Fu E (2013) Progress in the development of paper-based diagnostics for low-resource point-of-care settings. Bioanalysis 5:2821–2836
66. de Araujo WR, Paixão TRLC (2014) Fabrication of disposable electrochemical devices using silver ink and office paper. Analyst 139:2742–2747
67. Arena A, Donato N, Saitta G, Bonavita A, Rizzo G, Neri G (2010) Flexible ethanol sensors on glossy paper substrates operating at room temperature. Sensors Actuators B Chem 145:488–494

68. Murphy A, Gorey B, de Guzman K, Kelly N, Nesterenko EP, Morrin A (2015) Microfluidic paper analytical device for the chromatographic separation of ascorbic acid and dopamine. RSC Adv 5:93162–93169
69. Yoshihiro K, Hodges S, Cook BS, Zhang C, Abowd GD (2013) Instant inkjet circuits: lab-based inkjet printing to support rapid prototyping of ubicomp devices. Proceedings of the 2013 ACM International Joint Conference on Pervasive Ubiquitous Computing—UbiComp '13, p 363–372

Chapter 5
Printed Functionalities on Paper Substrates Towards Fulfilment of the ASSURED Criteria

Suzanne Smith, Dario Mager, Jan G. Korvink, and Kevin J. Land

Abstract Paper-based diagnostics usually refers to microfluidic and chemical diagnostic components implemented on paper substrates, with the aim of such devices being able to replace more costly microfluidic diagnostic platforms. Paper provides a model platform on which to develop low-cost diagnostics, as it has intrinsic fluidic properties in addition to being a low-cost substrate. For successful paper-based diagnostics to be realized, additional functionality needs to be integrated to enable paper-based diagnostics to fulfil other technical requirements. The recent technology push for printed electronics provides an opportunity to upgrade paper diagnostic devices with strongly augmented functionality, whilst maintaining low cost. This is a possible route to meeting the ASSURED criteria, and in this chapter we explore different options towards achieving this goal.

5.1 Introduction

The importance of developing point-of-care diagnostics that fulfil the ASSURED principles has been highlighted throughout this book. Paper-based microfluidics and sensors have been applied with some success to health diagnostics, environmental monitoring and food quality control [88, 109], but additional functional components are required to be integrated. Building blocks from the field of printed electronics such as sensors, processors, displays, connectivity and power have the potential to extend the functionality of these devices, and will be explored in this chapter.

S. Smith (✉) · K. J. Land
Council for Scientific and Industrial Research, Pretoria, South Africa
e-mail: ssmith@csir.co.za; kland@csir.co.za

D. Mager · J. G. Korvink
Karlsruhe Institute of Technology, Karlsruhe, Germany
e-mail: dario.mager@kit.edu; jan.korvink@kit.edu

K. J. Land (ed.), *Paper-based Diagnostics*,
https://doi.org/10.1007/978-3-319-96870-4_5

Table 5.1 Functional components for paper-based diagnostics to fulfil the ASSURED criteria

	A	S	S	U	R	E	D
Actuators and control modules				x		x	
Sensors		x			x	x	
Data processors and storage modules	x	x	x	x	x	x	
Read out and displays	x		x	x	x	x	
Connectivity modules	x			x	x	x	x
Energy storage components						x	x

These components build on the strengths of paper-based diagnostics such as affordability and specificity, by improving on the criteria that are often deficient in paper-based diagnostics

A fully ASSURED paper-based diagnostic solution would require a number of functional components to be integrated, including:

- *Actuators and control modules*—for user interaction with the device and execution of operational steps
- *Sensors*—for performing the required diagnostic and detection of additional environmental and external factors
- *Data processors and storage modules*—for obtaining a quantitative result and capturing/digitizing this data
- *Read-out and displays*—for conveying the result to the user directly and comprehensively
- *Connectivity modules*—for communication of the result from the test device to be stored and accessed in future as required
- *Energy storage components*—for powering the various functional components integrated as part of the paper-based diagnostic device
- *Integration*—for combining various functional building blocks into an automated, effective paper-based diagnostic solution

These modules contribute towards meeting the ASSURED criteria as summarized in Table 5.1, with details provided in the paragraphs to follow.

Paper-based health and environmental diagnostic tests usually require colorimetric, electrochemical or fluorescence based analysis methods [22]. These either rely on external equipment for analysis or on visual colorimetric read-out. Although the latter does not require equipment, the lack of a fully quantitative result, errors in visual analysis and documentation, as well as the skills and time required to perform the tests, limit the effectiveness of these types of solutions. Result read-out generally requires some form of external instrumentation to perform processing and communication of the result from the paper-based test. Common methods include the use of smartphones for either optical detection using colorimetric, fluorescence, or electrochemiluminescence techniques [102], or electrochemical detection [33]. A number of portable potentiostat systems have also been developed [19, 96, 176], and examples of paper-based diagnostics developed for use with external instrumentation have been showcased [61] (Fig. 5.1).

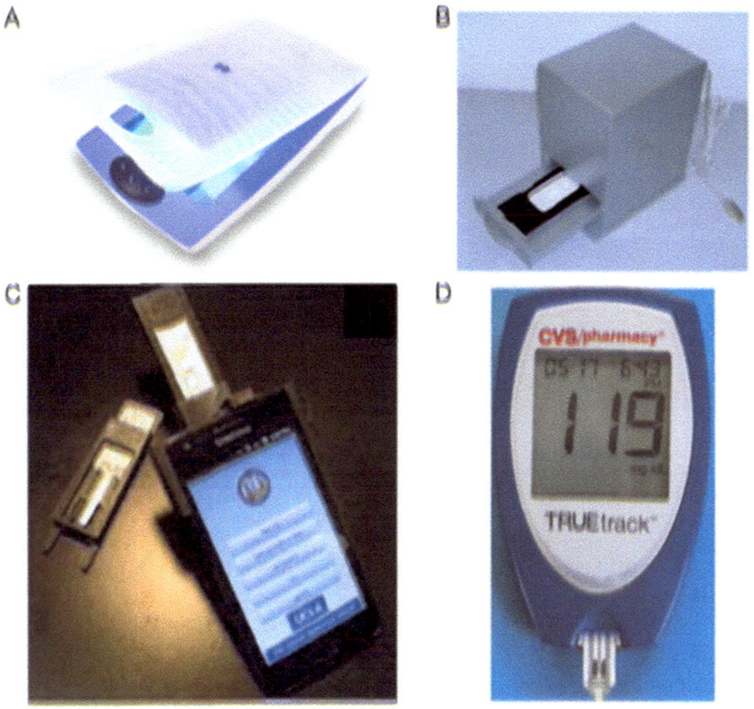

Fig. 5.1 Examples of portable instrumentation used for result read-out from paper-based diagnostic devices. Examples include (**a**) optical analysis by a scanner, (**b**) a portable reader, (**c**) a mobile phone and (**d**) a commercial electrochemistry-based reader. Reprinted from [61] with permission from Elsevier

These solutions add important computational capabilities to paper-based diagnostics, but can be susceptible to read-out errors as a result of user handling, and environmental factors such as lighting in the case of imaging of the result. In addition, low-resourced settings often have limited network infrastructure, intermittent power, and issues with theft, which can make instrumented solutions, particularly mobile-phone based ones—problematic to implement successfully. These limitations highlight the need for quantitative detection and read-out to be integrated, ideally on the diagnostic test strip itself, in an automated, standalone fashion to meet the equipment-free requirement.

The quantitative result obtained would need to be expressed to the user, possibly through a display or other feedback mechanism on the paper-based device, contributing towards the user-friendly requirement. A subsequent important consideration would be the capturing and storing of the test result, where, for example, significant challenges to be overcome for point-of-care testing in South African clinics lie in conveying and storing results and data [36, 110]. Connectivity of point-of-care diagnostic devices is thus crucial. Current manual result recording could be replaced by automated communication modules built into paper-based diagnostics, limiting user error and negating the need for external instrumentation.

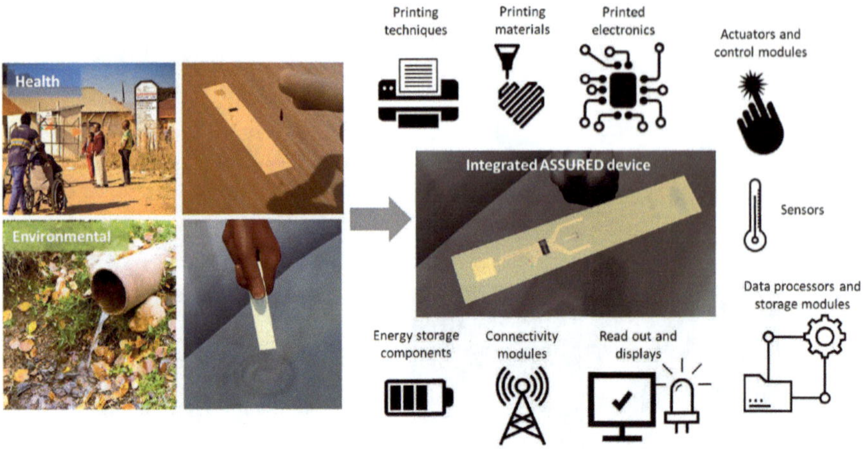

Fig. 5.2 Envisaged integrated, ASSURED paper-based diagnostic solutions for health and environmental testing, comprising a number of functional modules. Photo credits: CSIR, Pretoria, South Africa. Icon credits (clockwise from top left): Noun Project: Arafat Uddin, Bohdan Burmich, Michael Wohlwend, Magicon, dilakuscan, Creative Stall, DANT, Ralph Schmitzer, Vectors Market, DewDrops

Automated read-out, displays and communication units for paper-based diagnostics may require on-board power to drive the various added functionalities. This would also assist in fulfilment of the equipment-free requirement through realizing devices that are standalone with no external power sources required. Figure 5.2 shows a conceptual point-of-care diagnostic device with additional functional technologies integrated for health and environmental applications, encompassing the theme of this chapter.

5.2 Printing Techniques

Printing technologies are well established and date back to the invention of the printing press in the fifteenth century. Printing allows for high throughput production by utilizing roll-to-roll processes, making this a desirable platform on which to develop solutions for scale-up, including paper-based diagnostics.

A variety of printing techniques have been used in paper-based diagnostics, including wax printing of fluidic channels and screen printing of electrochemical sensors. Additional fabrication techniques for microfluidic paper-based devices that utilize printing technologies have been detailed and include inkjet printing and flexographic printing [133, 161]. Printing of electronics has seen extensive development in recent years, largely for flexible and wearable technologies and markets, and can be implemented using many different printing techniques [75], a number of which overlap with those used for paper-based microfluidics. Commonly used printing techniques for printed electronics onto flexible substrates include screen, inkjet, gravure, flexographic, off-set, slot-die, newer methods of transfer printing, as well as micro-contact and nanoimprinting. Low-cost printing solutions for printing copper tracks using a standard laser printer have also been showcased [174], and examples of printed electronics onto paper substrates using various printing techniques are summarized by Liang and Zou [92].

Patterning of paper and alternative flexible substrates for printing of integrated paper-based diagnostics has also been explored [13, 68, 98]. Technologies such as hot embossing—although not strictly a printing technique as no inks are used—also enable microfluidics and electronics to be integrated in flexible formats. Electronic structures and functional components are transferred from a stamp to a substrate. These so-called lab-on-a-foil processes have been successfully scaled up to continuous methods such as roll-to-roll [42] and have been applied to the low-cost production of highly integrated microfluidic devices [122]. These processing techniques can also be applied to paper-based devices [67]. Integrated solutions using all inkjet printing methods to realize electrical and sensing capabilities in addition to microfluidic functionality have also been explored [144].

Screen printing in both printed electronics and paper-based diagnostics has been a favourable trend as it is an established printing technology that allows for ease of printing onto different substrates and is applicable to both small and large scale manufacturing. Paper-based printed electrodes for biosensing applications are currently largely implemented using screen printing and many examples have been reviewed [137]. Wax screen printing also allows for fast, simple manufacture of paper-based microfluidic devices [133, 161], again highlighting the attractiveness of screen printing for fabrication of paper-based diagnostics with added functionality. Figure 5.3 illustrates low and high volume screen printing processes using manual and semi-automated sheet-to-sheet set-ups as well as large volume automated equipment for reel-to-reel processing. A number of commercial screen printers are available, including options from THIEME GmbH & Co. KG, Germany, the DEK printing solutions range from ASM Assembly Systems GmbH & Co. KG, Germany, and many other companies [63].

Functional components often require multilayer printing where a dielectric printable material is used as an intermediate layer between two conductive layers. This is useful for components such as high frequency antenna designs or capacitors. Alignment becomes important in these cases and printer technology requirements may be more sophisticated.

Combining of different printing technologies—for example wax printing and printed electronics—has yet to be explored fully, and integration challenges need to be overcome to implement both fluidic and electronic functionalities on paper optimally.

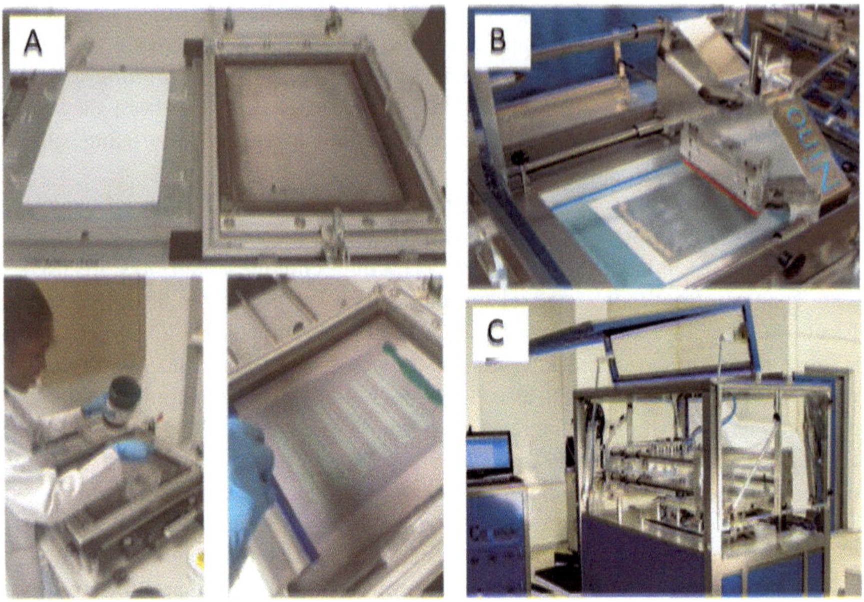

Fig. 5.3 (**a**) Manual sheet screen printing process, (**b**) semi-automated sheet by sheet screen printer (Nino—High Precision Screen Printer by Coruna Printed Electronics GmbH, Switzerland), and (**c**) automated screen printer set-up capable of reel-to-reel printing (Risto RtR—High Precision Screen Printer by Coruna Printed Electronics GmbH, Switzerland)

5.3 Printing Materials

Printed functionality can be achieved through the paper substrate itself, as well as the printable materials that are applied to the substrate. Modifications made to the paper substrate allow for heightened functionality of paper-based diagnostics [115]. In addition, printing of paper substrates themselves is currently being explored, for example by using printable nanocellulose pulp. From a printed electronics perspective, conductive inks, which typically consist of metallic particles such as silver, copper or carbon, are a foundational material. Carbon is generally less conductive than metals but lower in cost and often has favourable chemical properties.

Both inkjet and screen printable inks contain solvents and are in a fluid state when printed. These then need to be cured to remove the solvent and to solidify the ink to adhere to the substrate and make the ink conductive. Some conductive inks require UV or temperature curing, and many polymer-based inks cure at lower temperatures (around 150 °C), making them suitable for use on low cost substrates such as paper. Combining functional conductive elements successfully into cellulose substrates such as those used for paper-based diagnostics has been explored, but proves to be challenging [160]. Printed conductor materials can be used as electronic circuit tracks and components, electrodes and antennas. Transparent conductors are important for various printed functionalities, including solar cells, organic light emitting diodes (OLEDs) and touch screens [122], some of which will be discussed in more detail in Sect. 5.9: Read-out and displays on paper.

Additional printed materials for implementing added functionality on paper-based diagnostic devices include semi-conductors and dielectrics [75, 92], useful for realizing multi-layered structures and electronic components such as capacitors and transistors (see Sect. 5.5: Printed electronic components). Both inorganic and organic materials can be printed. In recent years, specialized functional inks have also become commercially available and include pressure and temperature sensitive inks. More details are provided in Sect. 5.7: Sensing on paper. The substrate choice for printing of functional components is also crucial for the successful implementation of the desired functionality [75], and a balance needs to be reached between the printed functionality achievable using paper substrates that are designed for use with paper-based diagnostics.

5.4 Hybrid Paper Circuits

Advances in printed electronics [106, 177] facilitates the development of paper-based diagnostics with built-in processing and automation. Functional electronics on paper require circuit layouts and tracks along with various electronic components. A straightforward approach involves utilizing existing packaged electronic components that can be mounted onto printed tracks on paper substrates—also known as hybrid printed electronics [9, 17, 58, 136]. Figure 5.4 shows examples of off-the-shelf coin cell batteries and surface mount LEDs and microcontrollers assembled onto printed, flexible tracks to create foldable, paper-based circuits. Figure 5.4 b illustrates how fluidics and electronics can be integrated into the same device, and these capabilities can be extended to form part of more complex paper-based diagnostic devices where folding or origami techniques are also utilized [22], as seen in Fig. 5.4c, d.

When combining existing electronic component packages such as integrated circuits or connectors with printed, flexible circuits on paper, different methods can be used to attach the electronic components to the paper, including low temperature soldering, conductive adhesives and paints, and crimping. Although these methods can effectively attach the components to the printed tracks, reliability remains an issue, as the rigidity of the components housed on flexible substrates can limit the flexibility of the device, and result in components detaching. Components in die formats or small, thin and flexible packaging can alleviate these challenges to a certain extent. However, integration remains a challenge as pick and placing of components and assembly remains a manual process in small quantities. Although automated processes for pick and place are available for larger production volumes, compatibility of components with the flexible substrates and the epoxy or adhesive needs to be studied to ensure effective implementation.

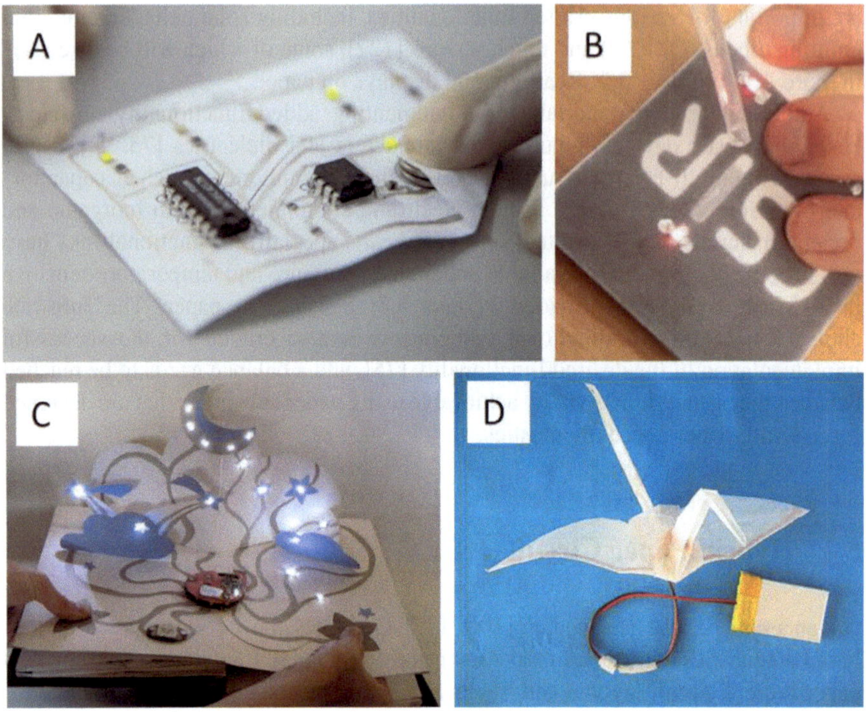

Fig. 5.4 Hybrid printed electronic circuits on paper-based substrates showing customizable designs and form factors. Photo credits: CSIR, Pretoria, South Africa (**a**, **b**), [125] (**c**, **d**)

5.5 Printed Electronic Components

Functional paper-based diagnostics will require a number of individual printed functional components to build up more complex electronic circuitry. This allows for interfacing of the various functional blocks and for processing capabilities to be implemented on the paper-based diagnostic devices, where both passive and active components will be required. A number of individual all-printed electronic components have been developed, including LEDs, capacitors, transistors, displays, solar cells and batteries, as well as RFID tags [23, 118]. These are often implemented on specialized printed electronics substrates, but examples have also been showcased on paper substrates [75]. Paper quality is an important consideration to achieve accurate geometrical dimensions of the components and to provide uniform surfaces on which to print specific heights of a particular material.

Figure 5.5 shows some examples of printed electronic components onto flexible plastic films. Gao et al. [47] also illustrated direct writing of resistors, capacitors and inductors onto different paper-based substrates including filter papers; substrates which are applicable to paper-based diagnostics. Temperature, bending and moisture, among other factors, can affect the performance and repeatability of these components [122].

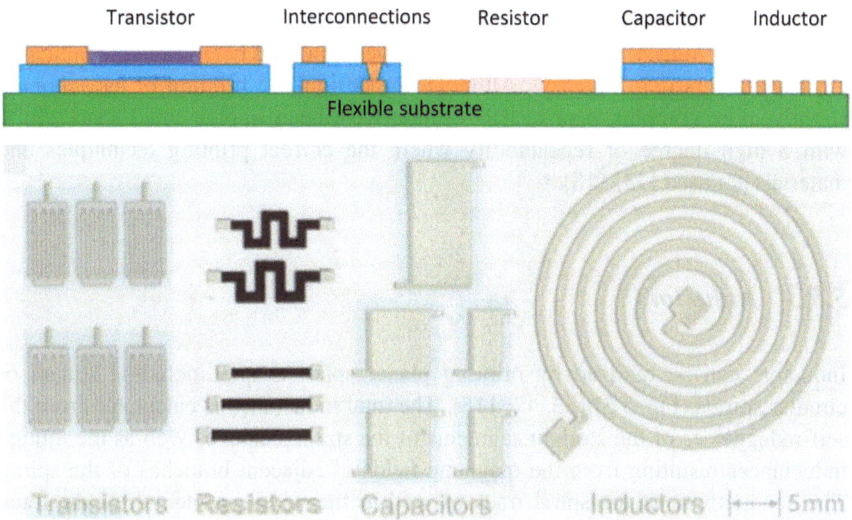

Fig. 5.5 Printed electronic components showing (a) structural designs, and (b) resulting printed components for resistors, capacitors, inductors and transistors. Reprinted from [23] with permission from Elsevier

5.5.1 *Resistors*

Resistors can be printed simply by varying the resistivity and geometrical dimensions of a conductive ink—typically carbon—that is printed over the ends of two metallic conductive track terminals. Design parameters for the resistance (R) include the length (L) and cross-sectional area (A) of the resistor which can be controlled through printing techniques, as well as the resistivity (ρ) ($\Omega\,\text{m}$) of the material used to print the resistor and are related by Eq. (5.1) as

$$R = \rho \frac{L}{A} \tag{5.1}$$

The simple design and manufacture process of printed resistors enables fully printable and repeatable resistive components to be realized. These can easily be integrated into existing printed circuitry and printed electronic solutions [23, 108].

5.5.2 *Capacitors*

Interdigitated capacitors can be implemented through lateral printed structures consisting of finger-like or comb designs. Capacitances achievable with these designs are limited to the picofarad (pF) range. In cases where greater capacitances are required, stacked capacitor structures can be utilized to achieve nanofarad (nF) capacitances. In these designs, a dielectric layer is printed between two conducting electrode layers (e.g. Fig. 5.5). Although the design and printing of capacitors is slightly more complex than resistors, these have been implemented successfully with a high degree of repeatability where the correct printing techniques and materials are used [23, 118].

5.5.3 Inductors

Inductors can be realized by printing planar spiral shapes including square or circular spirals (Fig. 5.5) [23, 47, 118]. The total inductance is calculated from the self-inductances of the straight segments of the spiral shape, as well as the mutual inductances resulting from the magnetic fields of adjacent branches of the spiral. The conductivity of the spiral track as well as the width and length of the track segments affects the resistance and thus the quality factor of the inductor. The spiral nature of printed inductors means that multilayer printing processes are required to provide connection from the inside of the spiral to the outside of the inductor shape. Good insulators are required to create this connection and to avoid leakage current, with a conductive track printed over the insulating layer to bridge the inner track to the outside. Fine printing resolution and alignment for multilayer printing are required for effective and repeatable printed inductors to be realized, but the design process is simple and robust.

5.5.4 Transistors

Transistors function as switches or amplifiers and are important in almost all electronic systems. Printed transistors typically consist of a gate, drain and source, with dielectric and semiconductor layers (Fig. 5.5). Paper-based inorganic and organic transistors have been developed, where the latter uses an organic semiconductor. Organic printed transistors are suited to mass production, require lower processing temperatures and may have cost benefits over printed inorganic transistor solutions. However, organic transistors face reliability issues with limited shelf life [70]. Cellulose-based paper is often used as a substrate or a gate dielectric for paper-based transistors, where the gate electrode can be printed on one side of the paper and the semiconductor and source/drain electrodes on the other. An overview of paper-based transistors is given by Liang 2016 [92]. Some examples include Inkjet-printed organic transistors [67, 82] and flexible, transparent transistors on nanopaper [179], as well as more recent flexible inorganic/organic hybrid transistors that showcase high performance [32]. The design and implementation of printed transistors makes use of multilayer printing processes and different materials, where more complex printing and alignment processes are required. However, as a result of

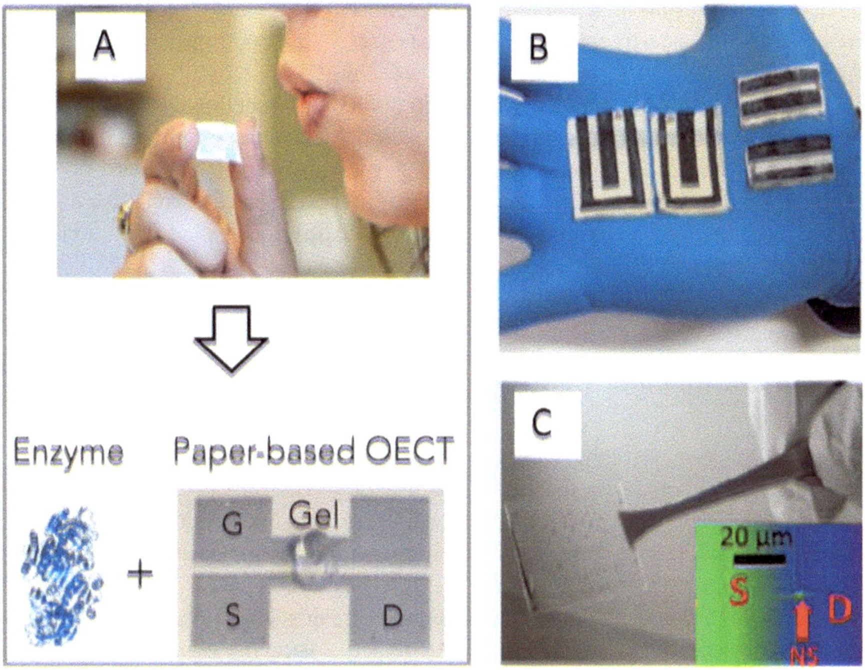

Fig. 5.6 (**a**) Printed organic electrochemical transistor as a disposable alcohol breathalyzer [20], (**b**) examples of transistors for wearable biosensors [55], and (**c**) high performance flexible inorganic/organic hybrid transistors, adapted with permission from [32]. Copyright 2017 American Chemical Society

the importance of transistors as functional components, this area is being explored extensively with fair progress in the repeatability and functionality of transistors implemented onto paper substrates.

Electrochemical transistors (ECTs) have been extensively investigated for their advantageous operation at low potentials, typically around 0.5–2 V, as the transistor is gated by an electrolyte. Printed batteries, supercapacitors and solar cells, which will be discussed in Sect. 5.9: Energy storage on paper—are thus suitable for powering these transistors. Paper-based ECTs have great potential for point-of-care diagnostics, with examples demonstrating simple, disposable alcohol breathalyzers [20], and wearable biosensors [55] (Fig. 5.6).

ECTs have also been developed by Acreo and Linköping University [2] where the gate insulator is a solid polyelectrolyte, resulting in a transistor that can operate at low voltages of less than 1 V, with short switching times of approximately 50 μs. This technology can be utilized in logic circuits, printed display driver circuits and active-matrix displays.

Individual printed electronic building blocks, from resistors to transistors, support the long-term goal of an all-printed paper-based diagnostic solution without the need for hybrid components and additional manufacturing steps. A fully integrated device would result in a paper-based diagnostic that would be lower in cost, robust, maintenance-free, and readily disposable—all contributing to becoming fully ASSURED.

Individual printed components can be combined into functional blocks to realize actuation, sensing, result processing and read-out, as well as communication of the result from the device to an external device or cloud server. On-board power is required on the device to drive the various components and to accomplish a stand-alone, integrated device with the potential to be ASSURED. These functionalities are briefly explored in the sections to follow.

5.6 Actuators and Control Modules on Paper

Integrated paper-based diagnostics will require some form of user interaction with the device. This could be to initiate the test or to perform potential operational steps on the device, while providing visible or tangible feedback to the user. Membrane switches have been widely implemented as a successful commercial printed electronics technology. These flexible membrane panels provide a simple solution using printed conductive tracks separated using adhesive spacers. When the button is pressed, the printed tracks make contact, closing the switch. Tactile spacer elements can be implemented to provide tangible feedback to the user. Membrane switch systems form part of most consumer and industrial electronic products that require human–machine interfacing.

These flexible printed technologies can be extended for use in paper-based diagnostics, and include demonstrations of paper buttons or switches [58]. Once the user has initiated a test the sample will be processed on the device, usually through multistep chemical or biological microfluidic processes. Timing, programming and other functional control in paper devices have been explored extensively and reviewed [22, 31], and include valves and different channel geometries for directing and manipulating fluid flow [103] (Fig. 5.7).

Paper-based actuators enable physical movement to be realized for paper structures. Electrically controlled paper actuators that undergo dimensional changes as a result of moisture variations have been demonstrated [58]. These can achieve different types of motion and could potentially be applied to paper-based diagnostics for fluid manipulation or where other physical movement would be required. Other examples include paper-based electroosmotic actuators [142] and paper actuators made from cellulose and hybrid materials [78]. High-performance paper actuators using standard printing paper and based on electrothermal activation have recently been demonstrated [7] (Fig. 5.8). The low voltage, power and temperature requirements would make these actuators well-suited for potential application to low-cost paper-based diagnostics.

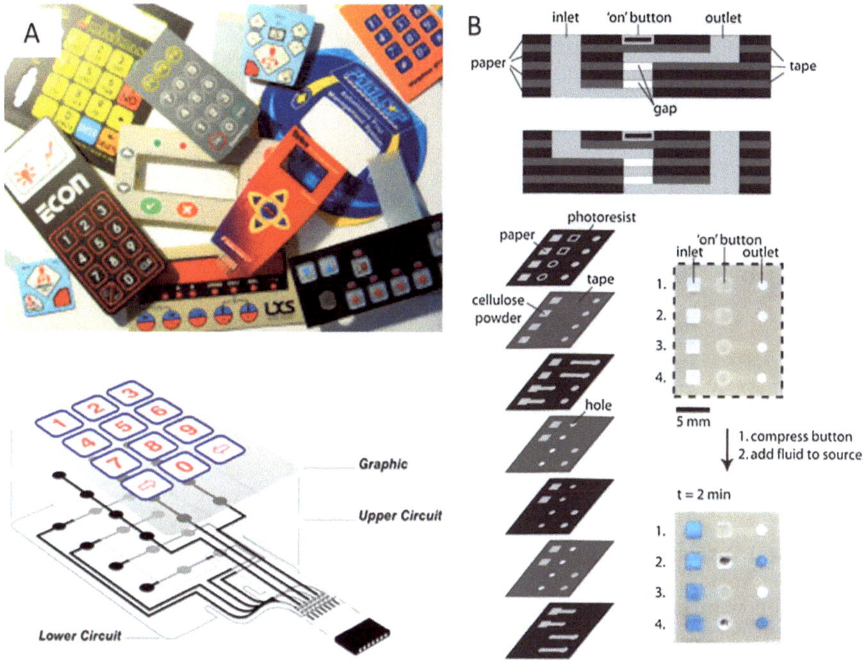

Fig. 5.7 Examples of (**a**) commercial membrane switch assemblies along with schematic. Reproduced with permission of Electronic Touch Systems (Pty) Ltd., South Africa, and (**b**) paper-based fluidic timing elements to start and stop fluidic operations. Reproduced in part from [103] with permission of The Royal Society of Chemistry

5.6.1 Heating on Paper

The combination of printed electronic elements with paper-based fluidics allows for heating components to be implemented with relative ease. Heaters have vast applications for chemical reactions and are thus highly relevant for integrated paper-based diagnostics development. In addition, heaters can facilitate sample concentration on paper through evaporation of liquids and can also support the implementation of valves on paper-based devices [104, 120] (Fig. 5.8).

Advances in materials for printed resistors allow for pastes with either positive or negative temperature coefficients to be printed. These inks can be used as self-regulating heating elements, where a specified temperature can be reached and maintained when a current is applied, without the need for additional control and regulating circuitry (Fig. 5.8). These functional positive or negative temperature coefficient inks could be useful for sample processing applications for integrated paper-based diagnostics where heating or temperature cycling at precise temperatures may be required. Carbon nanotubes have also been printed on paper and textiles to implement heating elements [41].

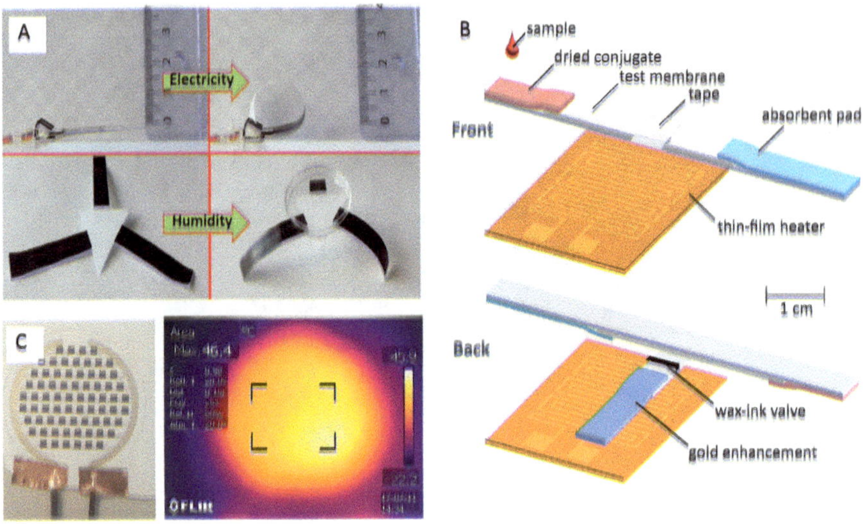

Fig. 5.8 (**a**) Paper-based actuators driven by electrical and humidity changes. Reprinted with permission from [7]. Copyright 2016 American Chemical Society, (**b**) wax valves implemented on lateral flow paper-based diagnostic strip and activated through a printed heater. Reproduced from [120] with permission of the Royal Society of Chemistry, and (**c**) printed heater implemented using silver and Loctite ECI 8000 series PTC inks from Henkel to realize a regulated temperature of approximately 46 °C. Image credit: CSIR, Pretoria, South Africa

Paper-based actuation and control can be achieved with relative ease in terms of mechanical actuation (buttons, switches, movement, etc.) and could be integrated with paper-based diagnostic devices. Fluidic actuation requires more precise implementations and integration considerations. Specialized functionality such as heating on paper is becoming more readily implementable with the development of sophisticated inks, and with the correct power supplied, can be integrated with paper-based diagnostics in a straightforward manner.

5.7 Sensors on Paper

Integrated sensing or detection techniques include colorimetric, electrochemical, electrical conductivity and chemiluminescent methods [88], many of which have been implemented on paper devices [18, 140]. A comprehensive summary of recent advances in printed sensors on foil substrates has been presented by Mattana and Briand [105], and includes both physical and chemical printed sensors such as temperature, pressure, pH and biosensors, and optical detection on flexible substrates. Advances in inkjet-printed micro-electro-mechanical systems (MEMS) have also been summarized [52]. This section briefly highlights printed sensing capabilities that have been implemented on paper that could allow for integration of automated detection of health or environmental parameters into paper-based diagnostic devices.

Printed resistive sensors form a large part of the printed sensing capabilities on paper substrates. Resistive sensors can be used in various applications that are based on strain or deflection sensing methods. Inkjet-printed resistive MEMS sensing devices manufactured from folded printing paper illustrate the simplicity with which these devices can be implemented on paper substrates, and thus readily applied to paper-based diagnostics without large cost implications [108]. Low-cost, easily manufacturable multisensory platforms on paper have recently been implemented, capable of detecting multiple parameters including temperature, humidity, proximity, pressure, pH and flow [113].

5.7.1 Temperature Sensors

Temperature is an important physical consideration that affects almost all measurements, and is thus an important parameter to monitor or detect on paper-based diagnostic devices. Most printed temperature sensors or thermometers are based on materials that vary in resistance as the temperature changes. Various examples of printed temperature sensors have been shown, using both inkjet printing and screen printing technologies to print resistors on substrates such as PET, other plastics, paper and fabric [105]. Temperature sensing and imaging on paper using temperature sensitive luminescence indicators have also been illustrated [80], where the paper device is laminated to prevent diffusion of oxygen. Commercial examples of printed temperature sensors have been showcased by Thinfilm in conjunction with Printed Sensor Technologies (PST) [123, 147] for example, where novel silicone inks allow for accurate temperature sensing to be implemented. Temperature sensors have been integrated with other printed electronic components into functional systems (see Sect. 5.12: Integrated paper-based devices), with high reliability, particularly on substrates such as PET.

5.7.2 Humidity Sensors

Humidity sensing, like temperature, is important, particularly for paper-based diagnostics, as moisture affects the functionality or outcome of the test. Relative humidity (RH) sensors can be printed in a planar fashion, with two or three layers of capacitor structures that detect a change in relative permittivity of the surrounding medium, and where an increase in capacitance indicates an increase in moisture content [122]. Different capacitor structures such as parallel plates, meshes and interdigitated designs can be utilized for humidity sensing. Successful printing of humidity sensors onto cardboard and recycled substrates has also been illustrated [111], and could potentially be applied to paper-based diagnostic substrates but requires further exploration. A paper-based electrical respiration sensor based on moisture content serves as a recent example with direct point-of-care applications [56].

5.7.3 Gas Sensors

Gas sensors on flexible substrates typically utilize resistive or capacitive sensors printed using specialized materials such as semi-conductive nanoparticle inks, insulation and solid electrolyte inks [122]. Highly sensitive gas sensors that have been developed include resistive nitrogen oxide gas sensors, and capacitive hydrogen gas sensors, as well as carbon monoxide and oxygen sensors [122]. Gas sensors have direct applications in health, environment and safety, with printed gas sensors thus highly relevant for paper-based point-of-care diagnostics. However, gas sensing remains a difficult problem, with paper-based substrates often not being uniform, introducing challenges in achieving accuracy and reliability.

5.7.4 Touch and Proximity Sensors

These sensors could be utilized to detect environmental surroundings or anomalies in the device handling, as well as for user interaction with a paper-based diagnostic device where user inputs are required. Available printed touch sensor technologies include capacitive, resistive, optical touch, acoustic wave, infrared and others, each suited to different applications [122]. The low cost of implementing resistive sensors makes this a favourable technology, followed by capacitive sensor technologies.

Printed resistive touch sensors are typically structured from two layers of patterned resistive substrates with an insulating spacer layer with an air gap in between. When the top sensor substrate is touched, the two conductive resistive substrates make contact, and current flows through the point of contact, enabling the touch position to be detected. Capacitive sensors can be used to sense touch or proximity of an object and these can be printed with relative ease. Either self-capacitance (single electrode) measurement systems or mutual capacitance (two electrode) measurement systems can be implemented. Touch and proximity sensors have previously been printed on paper-based substrates [74], showing promise for ease of implementation and integration with paper-based diagnostics.

5.7.5 Pressure Sensors

Printed and flexible pressure sensors could be useful in paper-based diagnostic devices to monitor or detect tampering or damage to the device. Simple graphite paper-based piezoresistive sensors have been realized [129]. Self-powered pressure sensors have also been demonstrated using triboelectric nanogenerators to convert mechanical energy to electrical energy [166] (see Sect. 5.11: Energy storage on paper—for more information). Wearable pressure sensors realized using gold nanowires on tissue paper sandwiched between polymer sheets have shown high

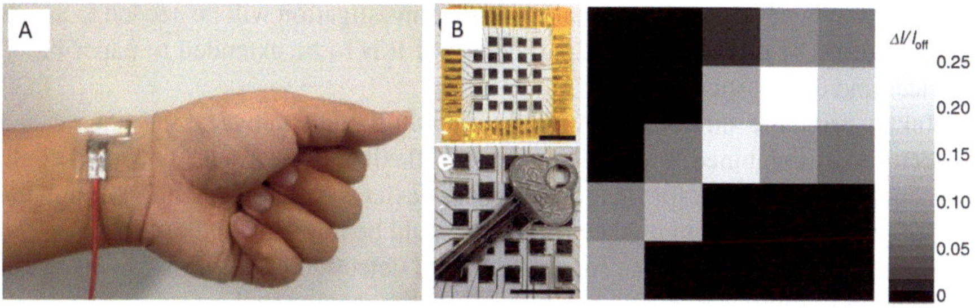

Fig. 5.9 (**a**) Wearable pressure sensors for pulse monitoring, and (**b**) pressure sensor matrix to detect weight distribution of objects. Reprinted by permission from Macmillan Publishers Ltd: Nature Communications [53], copyright 2014

sensitivity for blood pulse monitoring, as well as for detecting the weight distribution of small objects across an array of pressure sensors [53] (Fig. 5.9). Printed pressure sensitive inks are commercially available, and can be used to create free form pressure sensors across large areas in a matrix of sensors [75]. The resistance of the ink varies with the force applied to the printed ink surface, enabling different geometries and form factors to be implemented relatively simply. A common design approach is to use conductive interdigitated electrodes as a base layer with the pressure sensitive material printed over the top. Various form factors can be achieved as a result of the flexible and adaptable nature of these sensors, with many applications in healthcare. A number of commercially produced pressure sensors are available, showcasing the reliability of this technology, particularly on substrates such as PET, with potential for ease of adaption to paper-based diagnostics.

5.7.6 Light and Image Sensors

Printable light sensors or photodetectors have recently been developed and are typically realized by an organic (or less commonly inorganic) photoactive layer between two transparent electrodes [105]. Applications of printed photodetectors include industrial and medical large-area sensing, particularly for colour sensing, and could be highly relevant to colorimetric detection implemented on paper-based diagnostic devices towards an all printed ASSURED solution.

Photodetectors printed onto paper have been demonstrated [3], along with inkjet-printed photodiodes [94, 121], UV photodetectors and infrared sensors [52]. An overview of the components and integration of these into a functional flexible image sensor have also been showcased [66]. Organic photodiodes were integrated with organic thin film transistors in an active matrix to produce a 14 by 14 pixel image sensor on a plastic, flexible substrate. This technology is in the process of being commercialized by ISORG, France and Plastic Logic [65] towards a sensor with a 4 by 4 cm active area, 375 μm pitch and 94 by 95 pixel resolution [63]. As a fairly new area of printed functionality, further investigation will be needed to assess the reliability of this technology, particularly if it is to be extended to paper-based devices and diagnostics.

Inkjet printable microlens arrays for coupling light between the source and detector [52] combined with these printable, flexible image sensing arrays could offer potential solutions for imaging on paper devices. Colour detection through the use of photodetectors and image processing could be implemented directly onto the paper-based diagnostic device for sophisticated detection of colorimetric results or potentially in future for imaging of biological components such as cells or bacteria. Finger print imaging could be integrated using this technology to enable added security features as part of future envisaged paper-based diagnostic devices.

5.8 Data Processors and Storage Modules on Paper

In general, integrated sensing on paper-based devices requires analysis of the detected result, ideally using automated and accurate methods, without the user having to perform this task. This would allow for quantitative results to be obtained with fewer errors in read-out and without requiring user skill or training. Towards achieving this goal, printed components to implement processing or computational functions are required. Hybrid printed systems have been developed as discussed in Sect. 5.4. The use of die formats of integrated circuits has also been explored to provide a more flexible solution that has a smaller size and cost footprint. For fully printed and integrated solutions to be realized, functional blocks such as printed processing and memory components have been explored.

5.8.1 Integrated Circuits

Printed high gain amplifiers and digital-to-analogue converters (DACs) [23] consist of individual components that are built up to fully printed solutions for processing and amplification (Fig. 5.10). Three-dimensional printed integrated circuits were presented where a printed ring oscillator circuit was demonstrated [62]. Printed programmable logic circuits have also been implemented using transistor arrays [141]. Among the circuit functions and features demonstrated on the arrays were a ring oscillator, a D-type flip-flop memory element, a demultiplexer and a programmable array logic device. As a result, printed smart plastic systems made up of thin film transistors and printed electronic components to drive an electrophoretic display were realized. Medium-scale integrated circuits consisting of 100 transistors have been showcased [21], and more recently inkjet-printed thin film microprocessors have been realized [112], where the 8-bit thin-film microprocessor consists of a memory module that can be programmed after manufacture via inkjet printing. These developments show promise for complex printed integrated circuits to be

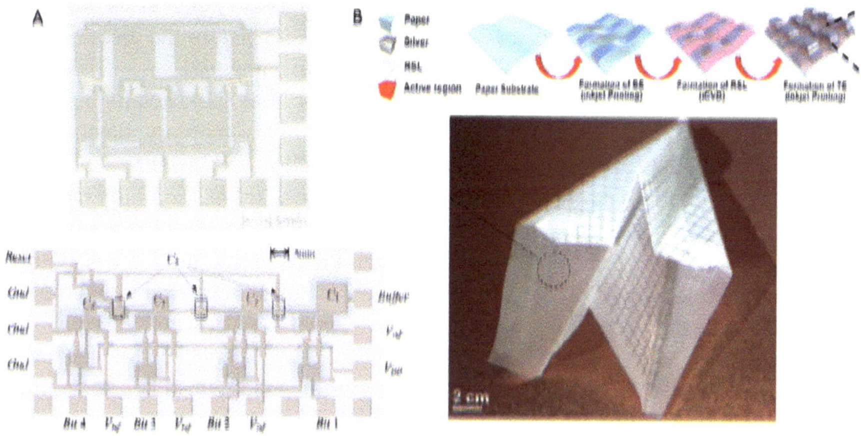

Fig. 5.10 (**a**) Screen printed amplifiers and 4-bit DAC modules on flexible substrates. Reprinted from [23] with permission from Elsevier, (**b**) paper-based substrate foldable memory, illustrating the printing of the various layers, including bottom electrode (BE), resistive switching layer (RSL) and top electrode (TE) [85]

produced in large volumes and become commercially implemented. Integration with paper-based diagnostics will need further investigation, where a balance between complexity and required functionality will need to be reached to keep costs low. The type of paper used could again pose limitations, where paper-based fluidics and electronics require different properties.

5.8.2 *Memory Components*

Data storage requirements are an important consideration in printed electronics generally, as well as for paper-based diagnostics, where sensor data logging and storing of user input data would typically be required. Data storage requirements could range from simple applications such as recording a threshold value to more complex logging applications where clock and timing functionalities would need to be included. Printed memory functionality has been demonstrated, typically using transistors, where a current needs to be applied to keep the stored memory state in a transistor.

Printed memristors show promise for implementing printed memory components, as these undergo resistance changes in response to a current applied. A positive current increases the resistance, and the resistance stays at a particular value until a negative current is applied, which decreases the resistance. A memristor thus only needs a current to change the state, not to keep it, as in the case of a transistor. Memristors as a memory component are thus more energy efficient and physically smaller than transistors. A memristor consists of two electrodes with a thin layer of memristive material between them. Inkjet printing has been used on plastic substrates to implement memristors effectively [150]. Typically the electrodes are made from a silver ink with a titanium oxide memristive material.

Write once read many (WORM) permanent memory has also been implemented in printed form to allow for more durable and complex memory functionality. WORM memory has been effectively developed by VTT [151], where the WORM memory bits are arranged as a one-dimensional array with one common electrode and a separate electrode for each bit [63]. Power requirements for these modules are typically in the range of 60 mW or less [122]. Inkjet printing and large scale roll-to-roll fabrication using gravure and flexography methods have been implemented successfully for this technology.

Some examples of memory on paper devices have also been demonstrated [85, 93] (Fig. 5.10). With advances in printing onto paper substrates to improve conductivity, foldability and disposability [155], printed paper-based data processing and storage systems are likely to evolve. These will play an important role in creating automated, integrated all-printed systems, examples of which are discussed in more detail in Sect. 5.12: Integrated paper-based devices.

With some commercially available as well as paper-implementation examples for memory components, this is a promising area of development. Reliability would need to be explored extensively to ensure that memory could effectively be implemented on paper-based diagnostic devices for read-out of results at a later stage.

5.9 Read-Out and Displays on Paper

Result read-out and feedback to the user are important for practical, automated point-of-care diagnostic devices to be realized effectively. Displays can be used to show the progress of the test being carried out, to display the result of the test, and to let the user know if the test was successful or not. Various flexible and printed display and lighting technologies continue to be developed, with newer paper-based implementations gaining attention.

Paper-based diagnostics can make use of displays that are built into the test itself through chemical reactions and biological processes. Examples include counting-based, timing-based, distance-based and text-based semi-quantitative displays or read-outs on paper-based diagnostic (Fig. 5.11).

Detection principles such as chemiluminescence, where light is generated based on a chemical reaction, have also successfully been showcased in paper-based diagnostics. This technique is advantageous as it does not require an excitation source, as in the case of fluorescence. Comprehensive reviews of read-out and display techniques implemented for paper-based diagnostics are available [162, 169].

Visual read-out approaches are useful for providing semi or fully quantitative results on the device. Potential limitations could arise from human error in interpreting the results in more complex systems, particularly where user training

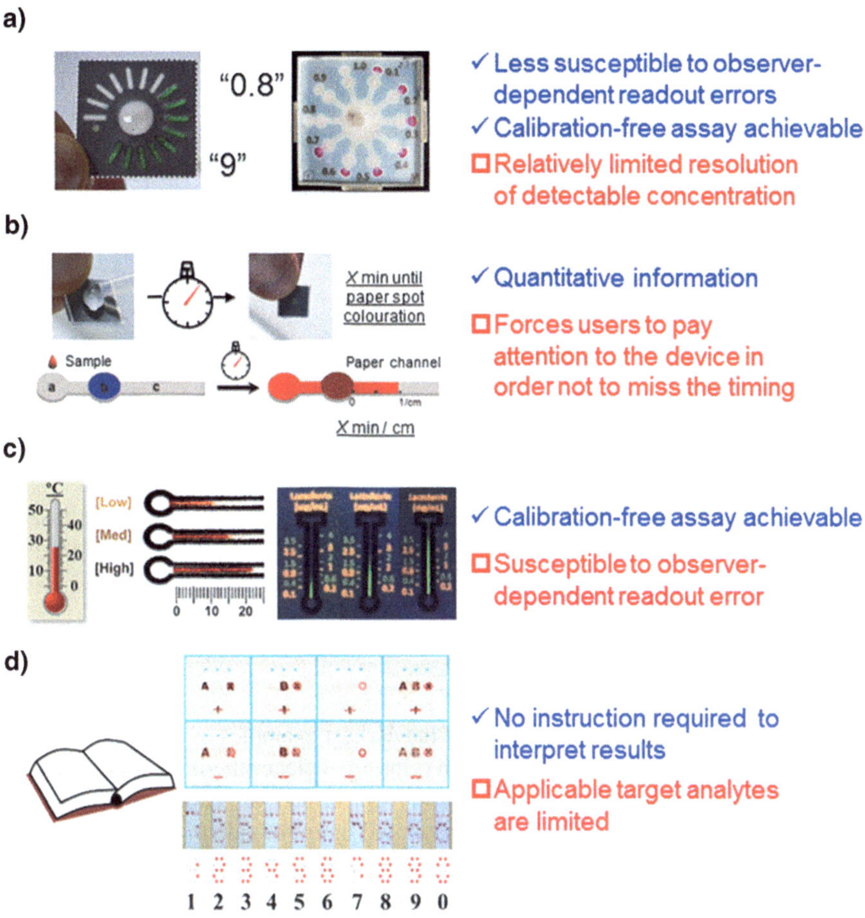

Fig. 5.11 Summary of currently implemented test result reporting systems on paper-based diagnostic devices, showing advantages and limitations of each. Reproduced from [162] with permission of The Royal Society of Chemistry. (**a**) Counting-based. (**b**) Timing-based. (**c**) Distance-based. (**d**) Text-based

is limited. By combining advances in printable displays and lighting technologies and paper-based diagnostics, read-out in an electronic format is achievable. In this digitized format, the result can be communicated, processed and stored without user interpretation required, while still providing visual feedback to the user. In addition to providing visual feedback to the user, printed display technologies can also be used to store encrypted information which can be decoded by an external system. Current printable displays include organic light emitting diodes (OLEDs), electrochromic, electroluminescent and electrophoretic displays [34, 40], some of which have been extended to paper-based systems.

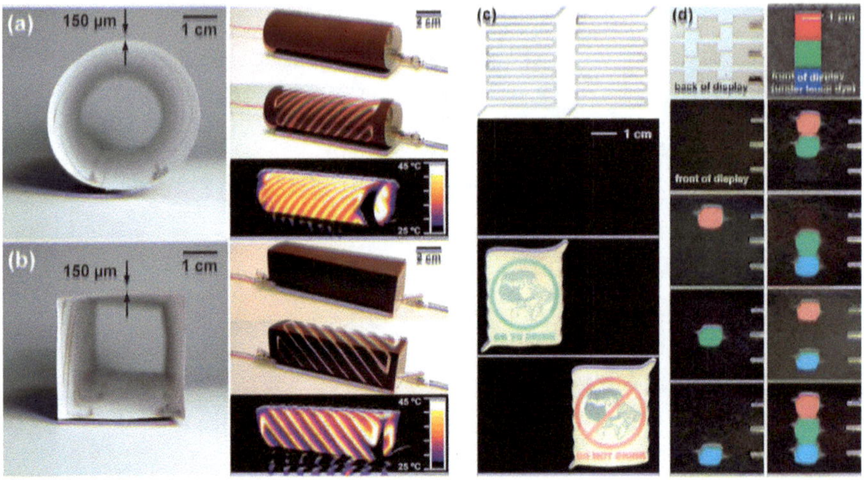

Fig. 5.12 Examples of thermochromic paper-based displays applied to flexible and foldable structures (**a**, **b**) and to paper-based diagnostics to provide a visual result (**c**, **d**). Reproduced from [135] with permission of The Royal Society of Chemistry

5.9.1 *Thermochromic Displays*

Thermochromic read-outs were some of the earliest examples of displays on paper-based platforms [135] (Fig. 5.12). Screen printing of electrothermochromic displays onto photopaper has also been effectively performed to realize seven-segment displays [134]. Thermochromic ink is patterned onto one side of a paper substrate, with conducting wires printed on the opposite side of the paper. When a current is passed through the wires on the paper, the thermochromic ink is heated and causes a colour change. Most thermochromic inks are reversible, meaning that when the temperature is lowered, the ink changes back to the original colour. Visual graphics are advantageous for these types of solutions over text, to maximise successful user interpretation, particularly in multilingual or illiterate environments. Thermochromic read-outs are reliable and effective and have already successfully been integrated into paper-based diagnostics [162].

5.9.2 *Electrochromic Displays*

Electrochromic displays operate in a similar fashion to thermochromic displays by converting an electrical signal into an optical read-out. Electrochromic materials change colour when they gain or lose an electron as a result of an applied current or potential. Reversing the polarity of the voltage causes the material to change back to its original colour. Electrochromic displays are advantageous as they are printable, flexible, transparent, have a low power consumption and do not require sophisticated encapsulation techniques [122].

These displays can be printed using standard techniques and organic components for both active and passive matrix addressed displays [37, 73], with an earlier example of an active matrix-based display implemented on coated paper by combining electrochemical transistors with electrochromic display cells into smart pixels [8]. Electrochromic read-out implemented in an integrated paper-based device with a paper battery and electrochemical sensor using Prussian blue has been showcased [95] and more recently to visually represent changes in resistive sensors [89]. Fully printable electrochromic materials are available, delivering commercial display and colour change solutions for smart cards, packaging and labels, for example, with potential future application to paper-based diagnostic solutions.

5.9.3 Light Emitting Electrochemical Cells

Light emitting electrochemical cells (LECs) provide an effective method of implementing flexible, lightweight display modules onto paper substrates. LECs are typically made up of an anode, cathode and active material which generates light when a voltage is applied. Low operational voltage requirements and air-stable electrodes make LECs power efficient with simplified encapsulation requirements [107]. These advantages in addition to the compatibility of LECs with paper substrates make this a promising emerging display technology for low-cost, paper-based point-of-care diagnostics.

Different types of paper have been utilized, including standard printing paper, where the fabrication of LECs did not require specialized processing [11]. A light emitting textile device using LEC technology has also been presented [83] where the anode and active material of an LEC is deposited by spray coating of a transparent fabric-based electrode, with potential applications to fibrous paper substrates such as those used in paper-based diagnostics. Recent advances have seen multi-coloured LECs being fabricated towards low-cost light sources with high luminance [97], which may be applicable to paper-based implementations in future.

5.9.4 Electroluminescent Displays

Electroluminescence enables a material to emit light in response to an electrical current or a strong electrical field. Electroluminescent displays consist of one or more layers of inorganic phosphor and insulator stacks sandwiched between two electrodes (an anode and cathode) that emits light when exposed to an electric field. Different types of electroluminescent displays include powder or thick-film displays thin-film displays, and hybrids of these. Screen printing is often used to manufacture powder ELDs, while thin-film ELDs utilize vapour deposition techniques. Powder electroluminescent displays have been successfully fabricated on different paper substrates, including glossy, sticker, newspaper and magazine papers [77].

Commercial electroluminescent displays can provide uniform light over relatively large areas and can be manufactured into thin structures using power efficiently [40], and can be used for lighting, displays and signage.

5.9.5 Electrochemiluminescent Displays

Electrochemiluminescence (ECL) is a derivative of chemiluminescence, where the signal is generated from an electrochemical reaction. Electrochemiluminescent read-out thus offers a higher degree of control by the user than chemiluminescence—which is purely reaction-based—but requires the ability to apply a potential. Examples of electrochemiluminescence implemented in paper-based diagnostics have been explored, and usually require external instrumentation to apply voltages, and to read out and quantify the emitted signal [162]. Various power supply methods, including thermal-powered, self-powered, bipolar and wireless techniques, and light detection methods including PMT, CCD, digital cameras and smartphones can be implemented as discussed in recent reviews [48, 54].

5.9.6 Organic Light Emitting Diodes (OLEDs)

Light emitting diodes (LEDs) are commonly used as displays and indicators in electronic systems. Organic LEDs (OLEDs) have been developed to create flexible and printable lighting and display solutions. A layer of organic material is sandwiched between two conductors (an anode and a cathode), which are then sandwiched between a transparent top and bottom plate. The organic layer of emissive electroluminescent enables bright red, green and blue light to be emitted in response to an electric current [122]. OLEDs differ from LEDs in that the light emitting material is organic, compared with the inorganic semiconductors in LEDs.

Different types of OLED technologies have been summarized [40], with distinctive shapes and sizes of displays achievable, making the technology customizable and also lending itself to low-cost and high-volume production. Inkjet printing has enabled mass production of flexible OLED devices through fast deposition of organic material without further patterning required. Encapsulation for protection from moisture and oxygen are important design considerations in the printing of OLED displays [122].

Although plastic-based OLEDs provide flexibility, paper is a lower cost alternative that could be explored and that would also be well-suited to paper-based point-of-care diagnostic solutions. The development of paper-based OLEDs has been limited as a result of the high sensitivity of OLED active materials requiring exact thicknesses, which can be difficult to achieve on paper [11]. Existing paper-based OLED technologies require sophisticated processing steps or specialized transparent papers, but have shown functional, flexible OLEDS made using standard printing paper [170] and transparent cellulose substrates [124].

5.9.7 *Electrophoretic Displays*

Electrophoretic materials do not produce light as in the case of OLEDs and ELDs, but rather make use of a process that forms letters or images by rearranging charged pigment particles using an electric field. Janus particles—nanoparticles that have surfaces with two or more different physical properties—are used to achieve this. Positively charged white and negatively charged black particles are suspended in liquid and allow for print quality resolution to be achieved as a result of the high reflectivity of the white state and high contrast [40]. Electrophoretic displays (EPDs) emulate traditional paper, with the most common type being e-paper, used in e-book reader products, and extended for use in watches and other mobile electronic displays with more colours now available [146].

As well as being a non-emissive technology that does not require backlighting, EPDs only consume power when the display is being updated. EPDs are commercially available and reliable and thus for point-of-care diagnostics for low-resourced settings this would be an advantageous display technology, utilizing very little power and providing high quality readability regardless of lighting conditions. EPDs have not been implemented directly onto paper substrates and thus integration of this technology with paper-based diagnostics has not yet been explored.

Display and read-out techniques presented in this section focus on technologies that are most relevant to paper-based devices aimed at low-cost diagnostics. A merge between fluidics and display technologies has been shown through electrowetting displays on paper [76] and electrofluidic displays [12]. Advances in organic light emitting transistors (OLETs) that combine the light emitting capabilities of an OLED with the switching capabilities of transistors show promise as a printable display applicable to various substrates, including paper [122].

5.10 Connectivity Modules on Paper

Once a measurement has been processed and some form of read-out has been implemented to provide feedback to the user, communication of the result from the disposable paper-based diagnostic device is required. This is important for effective result recording and filing of information. Result storage produces records that can be accessed by a medical doctor remotely towards all-connected solutions for standardized healthcare database systems. Wireless communication technologies are particularly advantageous for under-resourced settings, as they eliminate the need for physical connections to instrumentation and prevent contamination. Mobile phones have typically been used for result processing, read-out and connectivity for colorimetric tests, but can be problematic in resource-limited settings, as discussed in the Introduction.

Thorough reviews of portable connected instrumentation that have been developed for use with paper-based diagnostics are available [162] and include mobile phones, often with modifications or attachments, as well as electronics and optics housed in portable devices. Connections between paper devices and external instrumentation can be achieved through physical connectors and cables. Disposable USB connectors on paper have also been developed (e.g. Intellipaper [64]) and could be used for interfacing paper-based diagnostics with external equipment in a simple but effective manner. However, this section will focus on potential connectivity solutions that can form part of the paper-based device and move away from physical connections to external instrumentation. Wireless technologies are ideally suited for this application, with the printability of antennas offering additional advantages.

5.10.1 Wireless Communication Technologies

The evolution of the internet of things (IoT) has resulted in a drive to include connectivity between devices and systems [99]. Various wireless technologies could provide connectivity for paper-based diagnostic devices, with passive technologies such as radio-frequency identification (RFID) being attractive options as these are low-power, printable, low-cost, and thus ideally suited to paper-based diagnostic devices.

Existing wireless technology options include RFID, near field communication (NFC), Bluetooth, WiFi, ZigBee, Infrared, and cellular technologies such as 3 G and 4 G (e.g. see [116]). These technologies make use of different standards, frequency ranges and operational distance ranges. Bluetooth, WiFi and ZigBee operate at a frequency of 2.4 GHz, with ranges of tens to hundreds of metres, while cellular technologies operate in the frequency range of 450 MHz to 2.7 GHz, with ranges of a few km. However, most of these technologies are active, requiring power from an external source in order to operate. Passive short range communication technologies such as RFID and NFC are better suited from a low-cost, disposable device perspective, where power is not required on the device itself for the implementation of the communication module.

Power demands for wireless communication systems are dependent on the type of information being gathered by the sensor system, the interval at which the information is gathered, and the distance over which the information needs to be communicated [50]. Passive wireless technologies have been implemented for point-of-care diagnostic applications, and include implantable, wearable and hand-held wireless biosensors (WBSs) [51, 155]. As an interesting example, a flexible, passive wireless module for bacteria monitoring has been illustrated on tooth enamel [101].

RFID is typically used for tagging and tracking of goods in logistics and retail industries. NFC is a specialized subset of RFID technology, operating at 13.56 MHz with a range of 10 cm, and is applied in smartphones and tablets. For logistics, RFID has a number of advantages over existing identification techniques such as barcodes, including longer reading ranges, faster data transfer, and multiple, simultaneous tag reading capabilities. In addition, passive RFID tags draw energy from the electromagnetic field radiating from the reader when the tags are within the reader range, which is a favourable implementation since it does not require a battery or external power source, thus lowering the cost of the device. RFID technologies have also been highlighted as beneficial solutions for rural e-Health systems [26].

RFID technologies can be split into three main types according to the frequency of operation: low frequency (LF) RFID which operates in the frequency range of 120–150 kHz, high frequency (HF) RFID which operates at 13.56 MHz, and ultra high frequency (UHF) RFID which operates in the frequency range of 433–900 MHz. Ranges of operation for RFID technologies can be from approximately 10 to 40 m. Although the operational distances are shorter than for some of the other wireless technologies mentioned, these would be more than adequate for paper-based diagnostics which would need to be read within a clinic or hospital room, for example.

A number of wireless communication technologies have been established for healthcare systems to provide connectivity in hospitals and clinics. RFID has been used in hospitals for patient monitoring with some success [28]. Recently, privacy preservation in RFID systems for healthcare has also been pursued [128], to enable a secure, confidential method of data communication and storage, ideally suited to healthcare applications. Encryption of data is extremely important for handling sensitive information such as patient and health-related data, and RFID technologies allow for password protection to be implemented on tags, towards fulfilment of this requirement.

5.10.2 *Printed and Paper-Based Connectivity Modules*

The field of printed electronics, and specifically printed antennas, has seen rapid growth in recent years, enabling wireless communication systems to be implemented on flexible substrates [118]. This can be achieved using established, accessible, low-cost printing techniques such as screen printing, which lends itself to scale-up and mass manufacturability of printed devices. RFID is also well-suited to paper-based diagnostics as the RFID tag antennas are printable, with paper considered to be one of the best organic substrates for UHF and microwave applications [163].

As such, printing of antennas for wireless applications onto paper substrates using inkjet printing technology has been explored [1, 79, 132, 164]. Recently,

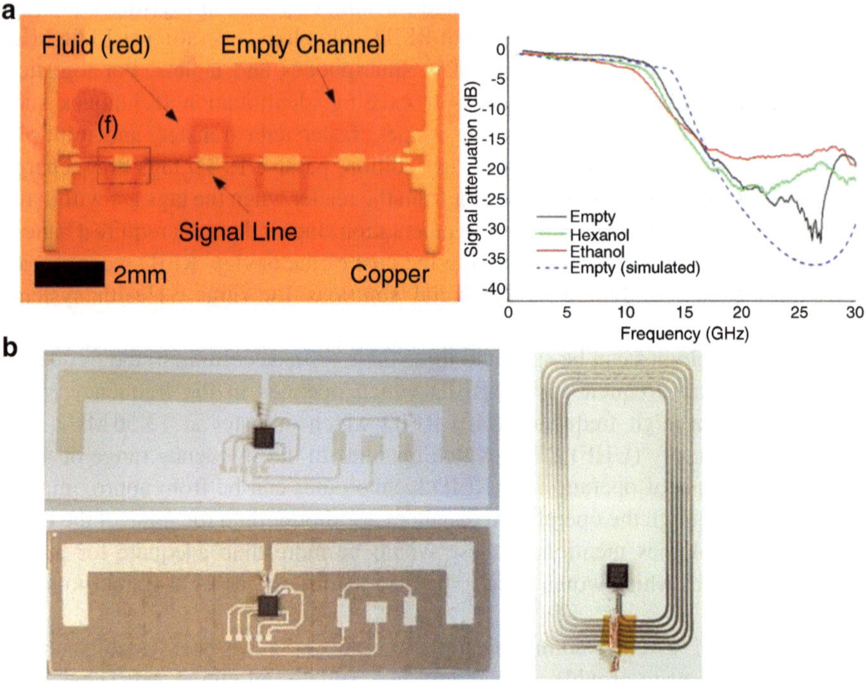

Fig. 5.13 (**a**) Combining of microfluidic channels and printed conductive tracks to realize reconfigurable antennas [144], and (**b**) screen printed UHF RFID sensing tags on low-cost substrates such as printing paper and cardboard as well as ink-jet printed NFC RFID tags [139]

printed UHF and NFC RFID antennas using silver and copper inks on low-cost paper and cardboard substrates have been investigated using both inkjet and screen printing techniques [29, 59, 71, 72, 138, 139] (Fig. 5.13). Combinations of RF, microfluidics and inkjet printing technologies to implement wireless sensing platforms that are activated by fluids in microfluidic channels have also been studied [30, 144] (Fig. 5.13).

There is scope for growth in RFID markets, particularly in areas of healthcare, with strong innovation in lowering the cost of manufacture and investigating different antenna and substrate materials [63], where printing technologies on paper-based devices plays a large role. Sensing RFID systems have evolved in recent years [175], adding to the value proposition of RFID technologies for connectivity of paper-based diagnostics, particularly since sensing and communication can be integrated into the device.

Some paper-based communication and sensing modules that have been implemented include patch antennas, UHF oscillators, mixers and frequency doublers [6]—which make up a variety of communication capabilities. In addition, paper-based microfluidics have been integrated with self-assembling RFID tags to create a self-powered biosensing platform [172]. Commercially available NFC RFID tags embedded into paper substrates are also available [10] for various applications including retail sectors.

5.11 Energy Storage Components on Paper

Added functionality for paper-based devices means that power will be required to drive the various components, which include sensing, result processing, read-out and communication. Power solutions that are stand-alone and can be integrated with the paper-based diagnostic device would be ideal, as under-resourced point-of-care locations and clinics may have limited or intermittent power available. Printed batteries, supercapacitors, fuel cells and solar cells are in the process of being developed, with some commercial products available and examples of integration with paper-based diagnostics showcased. The working principles of each of these energy storage devices and their differences have been documented well [159] and can be referred to for further information.

Paper is well suited for realizing energy storage devices, as cellulose substrates enable matrices of composites to be implemented, and the porous nature of paper is ideal as a separator between electrodes that can be filled with electrolyte [92]. Comprehensive reviews of printed and paper-based batteries and energy storage devices highlight the progress made in this field [46, 114, 117], and an extensive review by Yao et al. [168] summarizes paper-based electrodes for energy storage.

5.11.1 Electrochemical Batteries

Paper-based batteries can be loosely split into electrochemical and biofuel cells, with electrochemical batteries being the most commonly implemented on paper [92]. Electrochemical batteries are an established commercial technology, and the development of these into printed forms has accelerated in recent years. These batteries consist of two conductive electrodes typically made from metal or carbon with an electrolyte-filled paper in between to connect the two electrodes by a salt bridge or ion exchange membrane. Paper-based electrochemical batteries can be manufactured through deposition of electrodes onto paper substrates, and/or by introducing electrolytes into the paper (Fig. 5.14). Batteries can either be in sandwich or parallel formats, for which several printing techniques have been utilized [46]. Different types of papers have also been investigated, indicating that higher porosity papers tend to yield higher current densities.

Paper-based battery prototype form factors are typically in the region of $2\,cm^2$, providing average voltages of between 1 and 3 V. Paper separators containing electrolyte and can be constructed using origami/foldable or stacked paper designs with wax patterned wells for the various reagents [24, 148]. Paper-based batteries

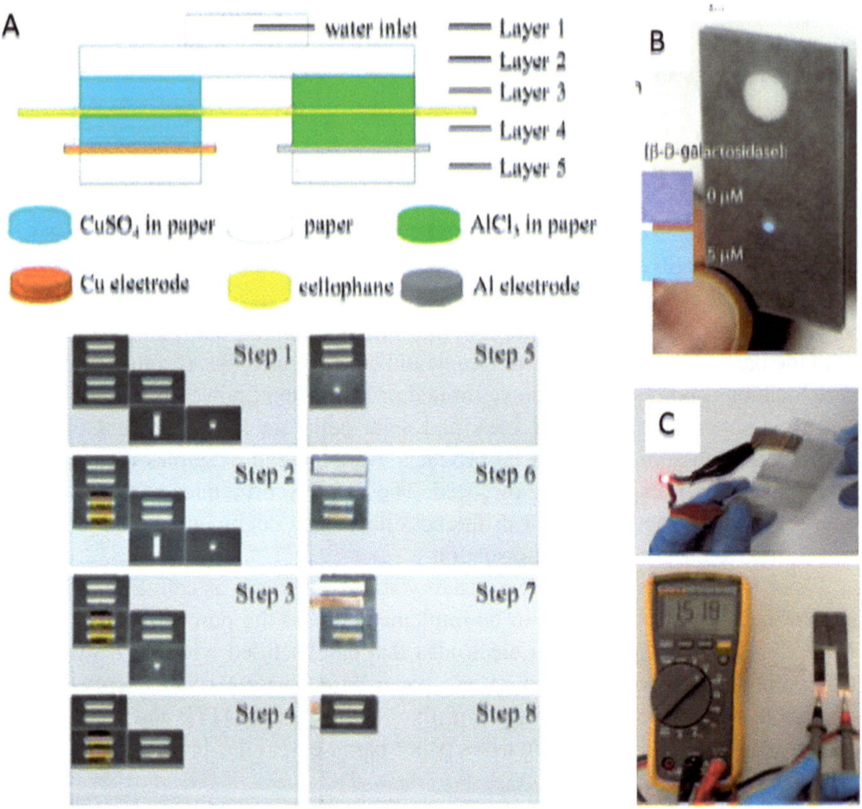

Fig. 5.14 Examples of paper-based electrochemical batteries: (**a**) multilayer, foldable wax printed designs. Reproduced in part from [24] with permission of The Royal Society of Chemistry, (**b**) an origami fluidic battery to illuminate a UV LED for fluorescence detection. Reproduced in part from [148] with permission of The Royal Society of Chemistry, and (**c**) an example of a low-cost printed primary battery with an open circuit voltage of 1.5 V, image credit: CSIR, Pretoria, South Africa

that are activated by introducing biofluids (e.g. urine) have also been implemented, and in one example, this has been integrated with electrochemical sensors and electrochromic read-out on a paper-based device [95].

Customization of printed primary batteries has been explored, both in terms of power and form factor requirements [158]. More recent developments in rechargeable polymer batteries have enabled thin, flexible, screen printed batteries to be realized, with potential applications in wearable electronics [145]. Commercially available primary printed batteries on substrates such as PET are reliable with specifications and shelf life comparable to traditional batteries. Development of fully paper-based batteries is still required to achieve comparable and commercially viable paper equivalent devices.

5.11.2 Lithium-Ion Batteries

Lithium ions are transported between the anode and cathode of lithium-ion batteries through an ion-conductive electrolyte that is electronically insulating. Paper-based lithium ion batteries typically make use of commercial papers that are coated with carbon nanotubes to create highly conductive, easily manufacturable battery components [60]. Lithium ion batteries are well suited to applications requiring high power and energy densities and show promise as flexible power sources [114].

Foldable lithium ion batteries have been developed using powders deposited onto current collectors consisting of paper coated with carbon nanotubes [25] (Fig. 5.15). A recent example demonstrates a thin (<250 μm), lightweight and flexible lithium-ion battery using graphite-based current collectors and with heightened conductivity [126]. Paper or paper-like lithium-ion batteries can generate power that is a few orders of magnitude larger than those in paper-based electrochemical or biofuel cells. These batteries also exhibit lengthy durability and are rechargeable. Drawbacks include the multiple layers of materials required, as well as the demanding encapsulation or packaging requirements for the battery, thus implying an increase in cost. Environmental friendliness of the battery components is also a shortcoming. Thus lithium-ion batteries may prove challenging to integrate with existing paper-based electronics and diagnostic devices, particularly in terms of cost and disposability, but could be well-suited to polymer-based flexible electronic applications [114].

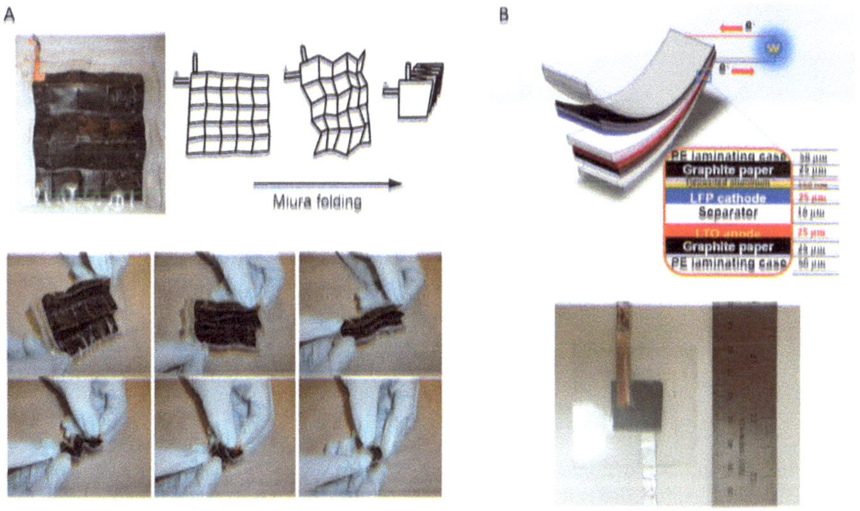

Fig. 5.15 Examples of paper-based lithium-ion batteries, including (**a**) a foldable design, reprinted with permission from [25]. Copyright 2013 American Chemical Society, and (**b**) a thin, flexible graphite-based and highly conductive prototype [126], ©2008 Canadian Science Publishing or its licensors. Reproduced with permission

5.11.3 Biofuel Cells

Biofuel cells utilize microorganisms or enzymes as catalysts to convert fuel (e.g. sugar, ethanol, methanol) and water into power, instead of using metal catalysts as in the case of electrochemical batteries [92]. These biofuel cells are thus advantageous in terms of cost, renewability and environmentally friendliness, as well as operating under mild pH and temperature conditions compared to standard fuel cells. Microbial fuel cells usually have long lifetimes, but lower power densities, while enzymatic biofuel cells have higher power densities but limited lifetimes [50, 114]. Typical sizes of some of the biofuel cell prototypes in literature are approximately 2–4 cm^2, with performances ranging from 10 to 90 $\mu AW/cm^2$ [114].

Fuel cells implemented on lateral flow devices have been investigated, illustrating the integration of power with standard paper-based diagnostic form factors for potential read-out and quantification of the result from the device [38]. Bacteria-driven paper-based biofuel batteries have been demonstrated [44, 84], capable of powering an LED for up to 30 min. Low-cost (less than $1 per device) paper-based microbial fuel cells utilizing a drop of sewage as fuel have been presented [43], with a more recent, stackable design developed [45] (Fig. 5.16).

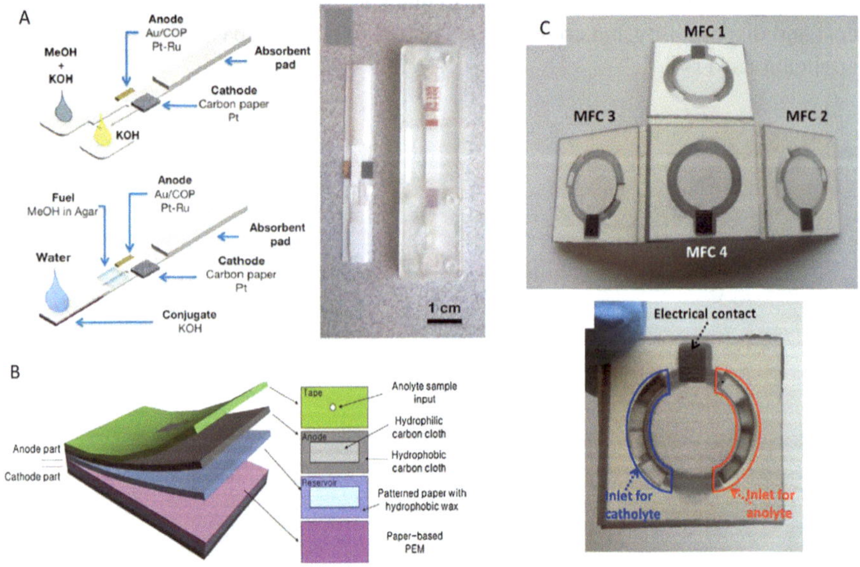

Fig. 5.16 (**a**) Paper-based biofuel cell integrated as part of lateral flow device [38], (**b**) schematic of paper-based microbial fuel cell paper layers, reprinted from [43] with permission from Elsevier, and (**c**) folded origami bacteria-powered batteries, reprinted from [45] with permission from Elsevier

5.11.4 Supercapacitors

Supercapacitors are electrochemical energy storage components, also referred to as ultracapacitors or electric double layer capacitors (EDLCs). Supercapacitors are similar to batteries in terms of design, manufacture and structural dimensions, consisting of two electrodes that are electrically isolated by a separator with an electrolyte. When a potential is applied, an electric field is formed over the interfaces between the electrode and electrolyte. The energy is stored in the electric field in a way that is comparable to conventional capacitors. Supercapacitors typically have higher power densities than batteries—a ten-fold increase compared with lithium ion batteries—and also have faster charge/discharge rates. They are considered to be more environmentally friendly and safer than batteries, but typically provide lower voltages. An in-depth review on the various aspects of supercapacitors provides more detail for the interested reader [157].

Similar to paper-based lithium-ion batteries, most paper-based supercapacitors are implemented on paper-like composites of cellulose, rather than standard unmodified paper, utilizing conductive carbon-based nanomaterials such as carbon nanotubes and graphene. Supercapacitors have been implemented on standard office papers with carbon nanotubes and ionic-liquid-based gel electrolytes [69], as well as on paper substrates using polyaniline-based supercapacitors [171].

A photoelectrical origami device with an internal chemiluminescent light source and an integrated paper supercapacitor was developed to collect and store generated photocurrents [49]. An all-solid-state supercapacitor was fabricated onto the origami device by drawing thin film graphite electrodes onto paper legs with a graphite pencil, and soaking the legs with gel electrolyte (Fig. 5.17).

Fully screen printed highly conductive electrodes on various flexible substrates for asymmetric supercapacitors have also been showcased [154], enabling simple, rapid and cost-effective manufacturing of supercapacitors. All printed supercapacitors utilizing conventional printing paper and a commercial desktop inkjet printer enabled various form factors and aesthetic designs to be realized [27] (Fig. 5.17). Supercapacitors fabricated using commercial pencils and office paper to realize interdigitated electrodes have also been demonstrated [35, 167]. Down et al. investigated both a pouch format with paper separators and liquid electrolyte, as well as a solid electrolyte, with high retention of capacitance observed after more than 5000 cycles.

Printed supercapacitors have found uses for memory back-ups, as well as in packaging and RFID systems [122]. The high current/power generation possible makes these devices an attractive solution, and paper-based implementations have gained attention, with the potential to be integrated with paper-based electronics and diagnostics.

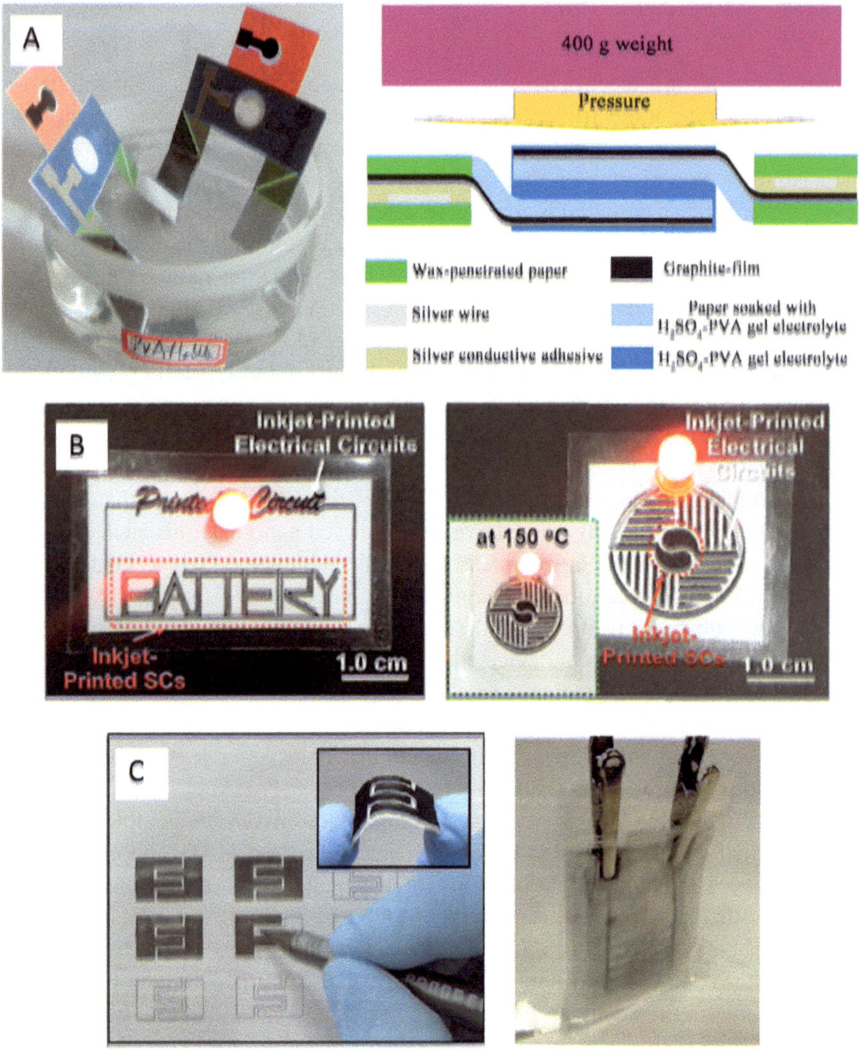

Fig. 5.17 Paper-based supercapacitor examples, including (**a**) origami-based device. Reprinted with permission from [49]. Copyright 2013 American Chemical Society, (**b**) supercapacitors made using desktop printer and conventional printing paper in different form factors. Reproduced from [27] with permission of The Royal Society of Chemistry, and (**c**) examples of pencil-drawn supercapacitors on standard paper substrates. Reprinted from [167] with permission from Elsevier and reproduced from [35] with permission of The Royal Society of Chemistry

5.11.5 Solar Cells

Solar photovoltaic systems generate direct current electricity from the sun's rays, with vast applications for generating electricity off-grid. Photovoltaic technology based on polymers has been gaining momentum, moving away from traditional silicon-based solar cells. Recent developments are focussing on thin-film inorganic, organic and dye-sensitized solar cell technologies [122]. Organic photovoltaic (OPV) technology offers a vast range of applications for flexible, lightweight and automated energy harvesting, from both solar and indoor illumination. OPV modules can be printed using roll-to-roll techniques with custom shapes and sizes achievable.

Paper-based photovoltaic cells have been investigated in recent years, including dye-sensitized solar cells on paper substrates [153] and printed photovoltaic cells on paper substrates that are recyclable [62]. Solar cells inserted into textiles using conventional weaving processes have also been demonstrated [173]. Novel papers with a balance between transparency and light scattering have been developed for effective paper-based solar cells to be realized [39]. Oxidative chemical vapour deposition (oCVD) was implemented by Gleason et al. for flexible, robust solar cell anodes to be realized on paper. A typical structure consists of an anode and a cathode, with a photoactive layer to enable solar cells to be printed on newspaper and standard printing paper [143].

An example of polymer solar cells with zinc-coating as the back contact and semi-transparent top electrodes demonstrated that the rough surface of the paper does not prohibit it from being used in high efficiency organic photovoltaics [87] (Fig. 5.18), and was illustrated in the case of cellulose-based electrodes and electrolytes implemented for dye-sensitized solar cells, with conversion efficiencies of up to 5% reported at simulated light intensities of 0.2 sun [14].

5.11.6 Nanogenerators

Nanogenerators are capable of harvesting energy from human movements, and can cleverly be implemented on paper using twistable and stretchable designs, making them an interesting area to explore for paper-based diagnostics. Origami-based nanogenerators and self-powered pressure sensors have been showcased and can achieve open circuit voltages of up to 20 V, short circuit currents of 2 μA, and peak power densities of 0.14 W/m^2 [166]. Paper-based nanogenerators for power and active sensors [178] illustrated that the action of turning the page of a book can generate up to 4.6 μA of current at a voltage of 2.4 V.

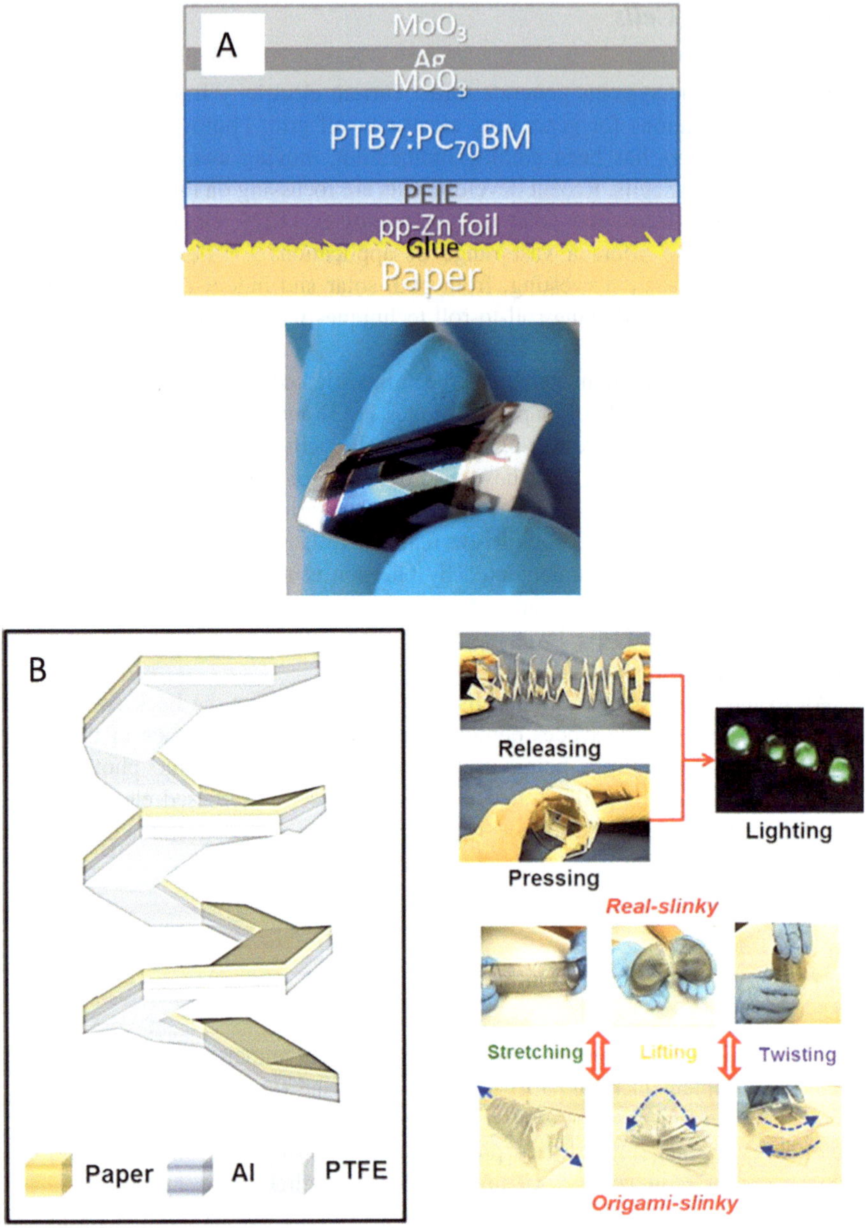

Fig. 5.18 (**a**) Structure and photo of polymer solar cells on paper substrates. Reprinted with permission from [87]. Copyright 2014 American Chemical Society. (**b**) Paper-based nanogenerator examples making use of slinky designs. Reprinted with permission from [166]. Copyright 2015 American Chemical Society

5.11.7 Energy Storage Components Summary

Commercially available printed batteries have been under development for a few decades, and are primary (i.e. non-rechargeable) batteries commonly using the well-known manganese dioxide/zinc battery chemistries. These batteries can be manufactured using screen printing technology with relatively inexpensive components [122]) and can be designed in such a way that scale-up is possible for roll-to-roll manufacture. Custom-made batteries can be developed, with main areas of development looking at specific power, thickness of battery, new materials, manufacturing methods, and serial connection of batteries. Printed rechargeable batteries with corresponding new or low cost materials are being explored.

Integrated power and sensing modules have been showcased for paper-based devices. Power demand versus the type of sensor information gathered and the distance over which communication must occur are all important considerations [50], as well as the duration for which the device should operate. Various integration considerations for energy storage devices are presented by Pan et al. [119].

In addition to the need for integrated power sources on paper-based diagnostics towards fulfilment of the ASSURED principles, the development of printed power sources has a number of advantages, including freedom of design, flexibility and environmentally friendly components, making printed power sources an attractive area of development not only for diagnostic and environmental monitoring solutions, but also for wearable technologies which have and continue to evolve at an exponential rate.

5.12 Integrated Paper-Based Devices

Thus far this chapter has explored various functional components that could be added to paper-based diagnostics towards the fulfilment of the ASSURED criteria. Merging of these components and integration with paper-based diagnostics is an important next step, and although measures have been taken to implement this, a number of challenges still remain. This section highlights some of the integrated paper-based devices that have been realized for enhanced functionality of paper-based diagnostics with the long-term goal of a fully ASSURED paper-based diagnostic solution in mind.

Advances in integration of paper-based devices are evident, as shown by the progression of lateral flow devices from using manual visual read-out to instrumented and automated read-out [100]. An overview on the convergence of microfluidics, electrochemistry and biosensors discusses the current state of the art and constraints in combining these technologies [127]. Additionally, developments in point-of-care diagnostic devices on flexible substrates have been summarized [155], and insights into the integration of printed electronics into the real world provide an interesting perspective on both current and future applications [16].

One of the primary integration obstacles lies in combining microfluidics and electronics on the same paper substrate, as each of these is suited to different paper characteristics [15, 57]. Integration of paper and polyester films to achieve optical and electrical sensing capabilities using antibodies and peptides has been investigated [131], as well as paper pre-treatment strategies to allow for high conductivity of dense, smooth copper tracks on paper [156]. Jenkins et al. [67] discussesd advanced printing technologies for both paper microfluidics and printed electronics towards integrated paper-based diagnostics.

Examples mentioned previously throughout the chapter that comprise integrated systems include connected paper-based sensing solutions with RFID wireless platforms for temperature sensing [152] and gas sensing [165]. Integration of printed passive components and a flexible battery for power electronics applications illustrates the pipeline of combining individual functional blocks in printed form [118]. Printed diodes have been combined with antennas and electrochromic displays to form an all-printed e-tag that is well-suited to IoT applications [130]. A paper-based electrochemical sensing platform with electrochromic read-out and integrated battery [95] as well as co-fabricated microfluidics and paper electronics and batteries [58] has also been developed. Interesting recent examples illustrate paper-based piezoresistive pressure sensors combined with visual read-out for measuring bandage compression [90], and infant birth weight [91] (Fig. 5.19).

Integrated electrochemical sensors with read-out in a flexible format has resulted in a so-called "printegrated" biosensor as part of a collaboration between Acreo AB and Linkoping University [15, 16, 149] (Fig. 5.19). The diagnostic device consists

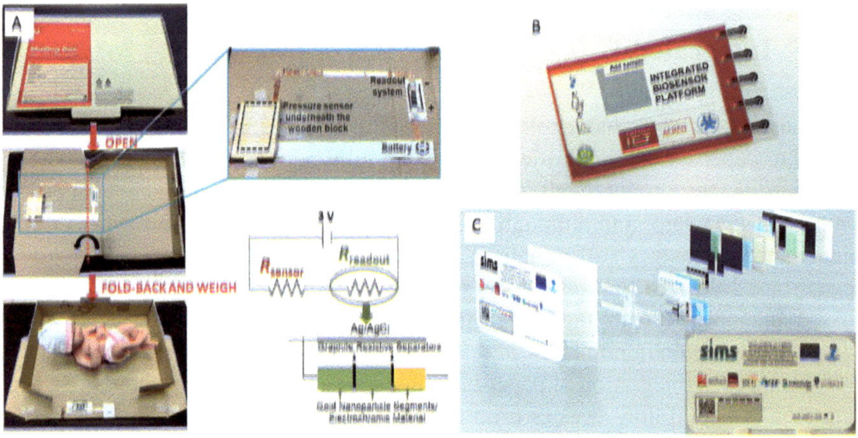

Fig. 5.19 Integrated printed solutions towards ASSURED diagnostics: (**a**) a "balance-in-a-box" birth weight system with paper-based read-out and a piezoresistive pressure sensor. Reproduced from [91] with permission of The Royal Society of Chemistry. (**b**) Printed electronic components combined to form smart, connected and flexible systems and printegrated biosensors [15]. (**c**) Integrated printed diagnostic device for measurement of hydrogen peroxide and total cholesterol. Reprinted with permission from [4]. Copyright 2017 American Chemical Society

of screen printed sensors, electrochromic display elements, a push button, printed battery and interconnects, and connectivity of the device via RFID or Bluetooth is in the process of being explored [15]. Lee et al. [86] also presented a device combining flexible electronics, microfluidics, and wireless connectivity to track micronutrient deficiencies. Paper-based microfluidics along with electronic instrumentation are housed in a thin, flexible format similar in size to a credit card. Measurements are wirelessly transferred to a mobile phone application for real time tracking of results. A recent example [4] shows the integration of printed electronics into a single device to measure hydrogen peroxide and total cholesterol. The disposable device consists of printed components including electrochemical sensors, an electrochromic display and printed battery. The system contains hybrid electronics, and is compatible with a mobile application for reading out results (Fig. 5.19).

Paper-based diagnostics that have been introduced to the market include solutions for liver function and nucleic acid testing that have been developed by Diagnostics For All (DFA), as well as text-reported blood typing by Haemokinesis and lateral flow devices for infectious diseases by INSiGHT (see [162] for more details).

The examples given emphasize the recent rapid developments made towards integrated printed and paper-based diagnostic solutions, and the variety of functionality that can be achieved. The technology building blocks to realize fully ASSURED paper-based diagnostics that are either in place or under development re-affirm the positive outlook for realizing effective, accessible paper-based diagnostics.

5.13 Challenges and Outlook

Although the technology advancements for enhanced paper-based diagnostics are progressing rapidly, unique challenges faced in under-resourced settings also need to be kept in mind. Case studies for low-cost diagnostics in the field have been conducted and illustrate the various hurdles faced [81]. Ultimately for these technologies to be successful in resource-limited settings and for the ASSURED principles to be attained, training, ease-of-use, and shelf life will need to be optimized. These design considerations will be important, and feedback from field tests and day-to-day handlers of the devices will be imperative for successful ASSURED diagnostics to be accomplished.

Scale-up and cost considerations for smart and connected paper-based diagnostics are important for these solutions to become viable and distributable. Roll-to-roll printing is an established technique and is being developed for mass production of printed electronics and functional systems. The pilot factory established by PrintoCent [122] has assembly processes in place for production of hybrid electronic devices, with the capability to handle bare dies with dimensions as small as 100 μm. Different components such as batteries, touch controls, displays and processing components can be integrated into the assembly line. Various conductive adhesives can be used, and in mold integration is possible for encapsulation and integration of microfluidics [5]. Mass production of paper-based microfluidics is a critical factor that is being explored. Lateral flow and dipstick tests already use established mass manufacturing techniques, but for newly developed paper-based diagnostics, scalability of microfluidic structure patterning, deposition of assay components, and assembly of multi-layered devices need to be studied [98, 162].

Substrates, inks, manufacturing processes, assembly of devices and fabrication time are all important cost factors to consider for the commercialization of paper-based diagnostics. Throughout the development process, costs will need to be kept low, while still providing functional diagnostic solutions, a trade-off that will continually need to be balanced as technologies and processes evolve. Printing technologies are excellent for these iterative improvements since they are highly flexible and show low tool changing costs.

References

1. Abutarboush HF, Shamim A (2012) Paper-based inkjet-printed tri-band u-slot monopole antenna for wireless applications. IEEE Antennas Wirel Propag Lett 11:1234–1237
2. Acreo (2016). www.acreo.se/expertise/printed-electronic-technologies
3. Aga RS, Lombardi JP, Bartsch CM, Heckman EM (2014) Performance of a printed photodetector on a paper substrate. IEEE Photon Technol Lett 26(3):305–308
4. Ahmadraji T, Gonzalez-Macia L, Ritvonen T, Willert A, Ylimaula S, Donaghy D, Tuurala S, Suhonen M, Smart D, Morrin A, Efremov V, Baumann RR, Raja M, Kemppainen A, Killard AJ (2017) Biomedical diagnostics enabled by integrated organic and printed electronics. Anal Chem 89(14):7447–7454. PMID: 28640594
5. Alajoki T, Koponen M, Tuomikoski M, Heikkinen M, Keränen A, Keränen K, Mäkinen JT, Aikio J, Rönkä K (2012) Hybrid in-mould integration for novel electrical and optical features in 3d plastic products. In: 2012 4th Electronic system-integration technology conference (ESTC), pp 1–6
6. Alimenti F, Mariotti C, Palazzi V, Virili M, Orecchini G, Mezzanotte P, Roselli L (2015) Communication and sensing circuits on cellulose. J Low Power Electron Appl 5(3):151–164
7. Amjadi M, Sitti M (2016) High-performance multiresponsive paper actuators. ACS Nano 10(11):10202–10210. PMID: 27744680
8. Andersson P, Nilsson D, Svensson P-O, Chen M, Malmström A, Remonen T, Kugler T, Berggren M (2002) Active matrix displays based on all-organic electrochemical smart pixels printed on paper. Adv Mater 14(20):1460–1464
9. Andersson HA, Manuilskiy A, Haller S, Hummelgård M, Sidén J, Hummelgård C, Olin H, Nilsson H-E (2014) Assembling surface mounted components on ink-jet printed double sided paper circuit board. Nanotechnology 25(9):094002
10. Arjowiggins Powercoat (2017). http://powercoatpaper.com/products/powercoat-alive/
11. Asadpoordarvish A, Sandström A, Larsen C, Bollström R, Toivakka M, Österbacka R, Edman L (2015) Light-emitting paper. Adv Funct Mater 25(21):3238–3245
12. Bai PF, Hayes RA, Jin M, Shui L, Yi ZC, Wang L, Zhang X, Zhou G (2014) Review of paper-like display technologies. Prog Electromagn Res 147:95–116
13. Ballerini DR, Li X, Shen W (2012) Patterned paper and alternative materials as substrates for low-cost microfluidic diagnostics. Microfluid Nanofluid 13(5):769–787
14. Bella F, Pugliese D, Zolin L, Gerbaldi C (2017) Paper-based quasi-solid dye-sensitized solar cells. Electrochim Acta 237:87–93

15. Beni V, Nilsson D, Arven P, Norberg P, Gustafsson G, Turner APF (2015) Printed electrochemical instruments for biosensors. ECS J Solid State Sci Technol 4(10):S3001–S3005
16. Berggren M, Simon DT, Nilsson D, Dyreklev P, Norberg P, Nordlinder S, Ersman PA, Gustafsson G, Jacob Wikner J, Hederén J, Hentzell H (2016) Browsing the real world using organic electronics, si-chips, and a human touch. Adv Mater 28(10):1911–1916
17. Bezuidenhout PH, Land KJ, Joubert T-H (2016) Integrating integrated circuit chips on paper substrates using inkjet printed electronics. In: 17th Annual conference of the rapid product development association of south africa (RAPDASA), 2–4 November 2016, Vaal University of Technology, Vanderbijlpark
18. Bezuidenhout P, Kumar S, Wiederoder M, Schoeman J, Land K, Joubert T-H (2017) The characterisation and design improvement of a paper-based E. coli impedimetric sensor. In: Proceedings of SPIE, fourth conference on sensors, MEMS, and electro-optic systems, 100360L, vol 10036, pp 10036–10036-6
19. Bezuidenhout P, Smith S, Land K, Joubert TH (2017) A low-cost potentiostat for point-of-need diagnostics. In: *2017 IEEE AFRICON*, pp 83–87
20. Bihar E, Deng Y, Miyake T, Saadaoui M, Malliaras GG, Rolandi M (2016) A disposable paper breathalyzer with an alcohol sensing organic electrochemical transistor. Sci Rep 6:27582
21. Cao Q, Kim H-S, Pimparkar N, Kulkarni JP, Wang C, Shim M, Roy K, Alam MA, Rogers JA (2008) Medium-scale carbon nanotube thin-film integrated circuits on flexible plastic substrates. Nature 454(7203):495–500
22. Cate DM, Adkins JA, Mettakoonpitak J, Henry CS (2015) Recent developments in paper-based microfluidic devices. Anal Chem 87(1):19–41
23. Chang J, Zhang X, Ge T, Zhou J (2014) Fully printed electronics on flexible substrates: high gain amplifiers and {DAC}. Org Electron 15(3):701–710
24. Chen S-S, Hu C-W, Yu I-F, Liao Y-C, Yang J-T (2014) Origami paper-based fluidic batteries for portable electrophoretic devices. Lab Chip 14:2124–2130
25. Cheng Q, Song Z, Ma T, Smith BB, Tang R, Yu H, Jiang H, Chan CK (2013) Folding paper-based lithium-ion batteries for higher areal energy densities. Nano Lett 13(10):4969–4974. PMID: 24059538
26. Chia S, Zalzala A, Zalzala L, Karim A (2013) Intelligent technologies for self-sustaining, RFID-based, rural e-health systems. IEEE Technol Soc Mag 32(1):36–43
27. Choi K-H, Yoo JT, Lee CK, Lee S-Y (2016) All-inkjet-printed, solid-state flexible supercapacitors on paper. Energy Environ Sci 9:2812–2821
28. Chowdhury B, Khosla R (2007) RFID-based hospital real-time patient management system. In: 6th IEEE/ACIS international conference on computer and information science (ICIS 2007), pp 363–368
29. Çiftçi T, Karaosmanoğlu B, Ergül Ö (2016) Low-cost inkjet antennas for RFID applications. IOP Conf Ser: Mater Sci Eng 120(1):012005
30. Cook BS, Cooper JR, Tentzeris MM (2013) An inkjet-printed microfluidic RFID-enabled platform for wireless lab-on-chip applications. IEEE Trans Microwave Theory Tech 61(12):4714–4723
31. Cunningham JC, DeGregory PR, Crooks RM (2016) New functionalities for paper-based sensors lead to simplified user operation, lower limits of detection, and new applications. Annu Rev Anal Chem 9(1):183–202. PMID: 27049635
32. Dahiya AS, Opoku C, Poulin-Vittrant G, Camara N, Daumont C, Barbagiovanni EG, Franzò G, Mirabella S, Alquier D (2017) Flexible organic/inorganic hybrid field-effect transistors with high performance and operational stability. ACS Appl Mater Interfaces 9(1):573–584. PMID: 28001361
33. Delaney JL, Doeven EH, Harsant AJ, Hogan CF (2013) Reprint of: use of a mobile phone for potentiostatic control with low cost paper-based microfluidic sensors. Anal Chim Acta 803:123–127
34. Deol RS, Choi HW, Singh M, Jabbour GE (2015) Printable displays and light sources for sensor applications: a review. IEEE Sensors J 15(6):3186–3195

35. Down MP, Foster CW, Ji X, Banks CE (2016) Pencil drawn paper based supercapacitors. RSC Adv 6:81130–81141
36. Engel N, Davids M, Blankvoort N, Pai NP, Dheda K, Pai M (2015) Compounding diagnostic delays: a qualitative study of point-of-care testing in south africa. Tropical Med Int Health 20(4):493–500
37. Ersman PA, Kawahara J, Berggren M (2013) Printed passive matrix addressed electrochromic displays. Org Electron 14(12):3371–3378
38. Esquivel JP, Del Campo FJ, de la Fuente JLG, Rojas S, Sabate N (2014) Microfluidic fuel cells on paper: meeting the power needs of next generation lateral flow devices. Energy Environ Sci 7:1744–1749
39. Fang Z, Zhu H, Yuan Y, Ha D, Zhu S, Preston C, Chen Q, Li Y, Han X, Lee S, Chen G, Li T, Munday J, Huang J, Hu L (2014) Novel nanostructured paper with ultrahigh transparency and ultrahigh haze for solar cells. Nano Lett 14(2):765–773. PMID: 24372201
40. Fernández MR, Casanova EZ, Alonso IG (2015) Review of display technologies focusing on power consumption. Sustainability 7(8):10854
41. Fischer T, Wetzold N, Elsner H, Kroll L, Hübler AC (2011) Carbon nanotube areas printed on textile and paper substrates. Nanomater Nanotechnol 1:3
42. Focke M, Kosse D, Muller C, Reinecke H, Zengerle R, von Stetten F (2010) Lab-on-a-foil: microfluidics on thin and flexible films. Lab Chip 10:1365–1386
43. Fraiwan A, Mukherjee S, Sundermier S, Lee H-S, Choi S (2013) A paper-based microbial fuel cell: instant battery for disposable diagnostic devices. Biosens Bioelectron 49(Suppl C):410–414
44. Fraiwan A, Dai C, Nguyen TH, Choi S (2014) A paper-based bacteria-powered battery having high power generation. In: The 9th IEEE international conference on nano/micro engineered and molecular systems (NEMS), pp 394–397
45. Fraiwan A, Choi S, stackable A (2016) Two-chambered, paper-based microbial fuel cell. Biosens Bioelectron 83:27–32
46. Gaikwad AM, Arias AC, Steingart DA (2015) Recent progress on printed flexible batteries: mechanical challenges, printing technologies, and future prospects. Energy Technol 3(4):305–328
47. Gao Y, Li H, Liu J (2013) Directly writing resistor, inductor and capacitor to composite functional circuits: a super-simple way for alternative electronics. PLOS ONE 8(8):1–8
48. Gao W, Saqib M, Qi L, Zhang W, Xu G (2017) Recent advances in electrochemiluminescence devices for point-of-care testing. Curr Opin Electrochem 3(1):4–10
49. Ge L, Wang P, Ge S, Li N, Yu J, Yan M, Huang J (2013) Photoelectrochemical lab-on-paper device based on an integrated paper supercapacitor and internal light source. Anal Chem 85(8):3961–3970
50. Gellett W, Kesmez M, Schumacher J, Akers N, Minteer SD (2010) Biofuel cells for portable power. Electroanalysis 22(7–8):727–731
51. Ghafar-Zadeh E (2015) Wireless integrated biosensors for point-of-care diagnostic applications. Sensors 15(2):3236–3261
52. Gih-Keong L, Shrestha M (2017) Ink-jet printing of micro-elelectro-mechanical systems (mems). Micromachines 8(6):1–19
53. Gong S, Schwalb WH, Wang Y, Chen Y, Tang Y, Si KJ, Shirinzadeh B, Cheng W (2014) A wearable and highly sensitive pressure sensor with ultrathin gold nanowires. Nat Commun 5:1–8
54. Gross EM, Durant HE, Hipp KN, Lai RY (2017) Electrochemiluminescence detection in paper-based and other inexpensive microfluidic devices. Chem Electro Chem 4(7):1594–1603
55. Gualandi I, Marzocchi M, Achilli A, Cavedale D, Bonfiglio A, Fraboni B (2016) Textile organic electrochemical transistors as a platform for wearable biosensors. Sci Rep 6:33637
56. Guder F, Ainla A, Redston J, Mosadegh B, Glavan A, Martin TJ, Whitesides GM (2016) Paper-based electrical respiration sensor. Angew Chem Int Ed 55(19):5727–5732
57. Hamedi MM, Ainla A, Güder F, Christodouleas DC, Fernández-Abedul MT, Whitesides GM (2016) Integrating electronics and microfluidics on paper. Adv Mater 28(25):5054–5063

58. Hamedi MM, Campbell VE, Rothemund P, Güder F, Christodouleas DC, Bloch J-F, Whitesides GM (2016) Electrically activated paper actuators. Adv Funct Mater 26(15):2446–2453
59. He H, Sydänheimo L, Virkki J, Ukkonen L (2016) Experimental study on inkjet-printed passive uhf RFID tags on versatile paper-based substrates. Int J Antennas Propag 2016. Article ID 9265159
60. Hu L, Choi JW, Yang Y, Jeong S, La Mantia F, Cui L-F, Cui Y (2009) Highly conductive paper for energy-storage devices. Proc Natl Acad Sci 106(51):21490–21494
61. Hu J, Wang S, Wang L, Li F, Pingguan-Murphy B, Lu TJ, Xu F (2014) Advances in paper-based point-of-care diagnostics. Biosens Bioelectron 54:585–597
62. Hubler A, Trnovec B, Zillger T, Ali M, Wetzold N, Mingebach M, Wagenpfahl A, Deibel C, Dyakonov V (2011) Printed paper photovoltaic cells. Adv Energy Mater 1(6):1018–1022
63. IDTechEx Research (2016). www.idtechex.com
64. intelliPaper (2017). https://www.intellipaper.info/
65. ISORG (2017). http://www.isorg.fr/
66. Jacob S, Benwadih M, Bablet J, Charbonneau M, Aliane A, Plihon A, Revaux A (2015) Large area sensing surfaces: flexible organic printed interfacing circuits and sensors. In: 2015 IEEE international electron devices meeting (IEDM), pp 19.5.1–19.5.4
67. Jenkins G, Wang Y, Xie YL, Wu Q, Huang W, Wang L, Yang X (2015) Printed electronics integrated with paper-based microfluidics: new methodologies for next-generation health care. Microfluid Nanofluid 19(2):251–261
68. Jiang X, Fan ZH (2016) Fabrication and operation of paper-based analytical devices. Annu Rev Anal Chem 9:203–222
69. Kang YJ, Chung H, Han C-H, Kim W (2012) All-solid-state flexible supercapacitors based on papers coated with carbon nanotubes and ionic-liquid-based gel electrolytes. Nanotechnology 23:065401
70. Kaushik BK, Kumar B, Negi YS, Mittal P (2012) Prospects and limitations of organic thin film transistors (OTFTs). Adv Comp Sci Eng Appl 167:125–139
71. Kavcic U, Pivar M, Dokic M, Gregor-Svetec D, Pavlovic L, Muck T (2014) Uhf RFID tags with printed antennas on recycled papers and cardboards. Mater Technol 48(2):261–267
72. Kavcic U, Maek M, Muck T (2015) Ultra-high frequency radio frequency identification tag antennas printed directly onto cardboard used for the manufacture of pharmaceutical packaging. J Imaging Sci Technol 59(5):50504-1–50504-8
73. Kawahara J, Andersson Ersman P, Nilsson D, Katoh K, Nakata Y, Sandberg M, Nilsson M, Gustafsson G, Berggren M (2013) Flexible active matrix addressed displays manufactured by printing and coating techniques. J Polym Sci B Polym Phys 51(4):265–271
74. Kawahara Y, Hodges S, Gong N-W, Olberding S, Steimle J (2014) Building functional pervasive computing prototypes using conductive inkjet printing. Pervasive Comput 13:30–38
75. Khan S, Lorenzelli L, Dahiya RS (2015) Technologies for printing sensors and electronics over large flexible substrates: a review. IEEE Sensors J 15(6):3164–3185
76. Kim DY, Steckl AJ (2010) Electrowetting on paper for electronic paper display. ACS Appl Mater Interfaces 2(11):3318–3323. PMID: 20973510
77. Kim JY, Park SH, Jeong T, Bae MJ, Song S, Lee J, Han IT, Jung D, Yu S (2010) Paper as a substrate for inorganic powder electroluminescence devices. IEEE Trans Electron Devices 57(6):1470–1474
78. Kim J, Yun S, Mahadeva SK, Yun K, Yang SY, Maniruzzaman M (2010) Paper actuators made with cellulose and hybrid materials. Sensors 10(3):1473–1485
79. Kisic M, Dakic B, Damnjanovic M, Menicanin A, Blaz N, Zivanov L (2013) Design and simulation of 13.56 MHz RFID tag in ink-jet printing technology. In: 2013 36th international spring seminar on electronics technology (ISSE), pp 263–267
80. Koren K, Kühl M (2015) A simple laminated paper-based sensor for temperature sensing and imaging. Sensors Actuators B Chem 210(Suppl C):124–128
81. Kumar AA, Hennek JW, Smith BS, Kumar S, Beattie P, Jain S, Rolland JP, Stossel TP, Chunda-Liyoka C, Whitesides GM (2015) From the bench to the field in low-cost diagnostics: two case studies. Angew Chem Int Ed 54(20):5836–5853

82. Kwon Y-J, Park YD, Lee WH (2016) Inkjet-printed organic transistors based on organic semiconductor/insulating polymer blends. Materials 9(8):650
83. Lanz T, Sandström A, Tang S, Chabrecek P, Sonderegger U, Edman L (2016) A light-emission textile device: conformal spray-sintering of a woven fabric electrode. Flex Print Electron 1(2):025004
84. Lee H, Choi S (2015) An origami paper-based bacteria-powered battery. Nano Energy 15:549–557
85. Lee B-H, Lee D-I, Bae H, Seong H, Jeon S-B, Seol M-L, Han J-W, Meyyappan M, Im S-G, Choi Y-K (2016) Foldable and disposable memory on paper. Sci Rep 6:38389
86. Lee S, Aranyosi AJ, Wong MD, Hong JH, Lowe J, Chan C, Garlock D, Shaw S, Beattie PD, Kratochvil Z, Kubasti N, Seagers K, Ghaffari R, Swanson CD (2016) Flexible opto-electronics enabled microfluidics systems with cloud connectivity for point-of-care micronutrient analysis. Biosens Bioelectron 78:290–299
87. Leonat L, White MS, Głowacki ED, Scharber MC, Zillger T, Rühling J, Hübler A, Sariciftci NS (2014) 4% efficient polymer solar cells on paper substrates. J Phys Chem C 118(30):16813–16817
88. Liana DD, Raguse B, Gooding JJ, Chow E (2012) Recent advances in paper-based sensors. Sensors (Switzerland) 12(9):11505–11526
89. Liana DD, Raguse B, Gooding JJ, Chow E (2015) Toward paper-based sensors: turning electrical signals into an optical readout system. ACS Appl Mater Interfaces 7(34):19201–19209. PMID: 26329490
90. Liana DD, Raguse B, Gooding JJ, Chow E (2016) An integrated paper-based readout system and piezoresistive pressure sensor for measuring bandage compression. Adv Mater Technol 1(9):1600143
91. Liana DD, Raguse B, Gooding JJ, Chow E (2017) A balance-in-a-box: an integrated paper-based weighing balance for infant birth weight determination. Anal Methods 9:66–75
92. Liang T, Zou X, Mazzeo AD (2016) A flexible future for paper-based electronics. In: Micro- and nanotechnology sensors, systems, and applications VIII, proceedings of SPIE, vol 9836, p 98361D-1
93. Lien DH, Kuo ZK, Huang TH, Liao YC, Lee SC, He H (2014) Paper memory by all printing technology. In: 2014 symposium on VLSI technology (VLSI-Technology): digest of technical papers, pp 1–2
94. Lilliu S, Boeberl M, Sramek M, Tedde SF, Macdonald JE, Hayden O (2011) Inkjet-printed organic photodiodes. Thin Solid Films 520(1):610–615
95. Liu H, Crooks RM (2012) Paper-based electrochemical sensing platform with integral battery and electrochromic read-out. Anal Chem 84(5):2528–2532
96. Liu B, Du D, Hua X, Yu X-Y, Lin Y (2014) Paper-based electrochemical biosensors: from test strips to paper-based microfluidics. Electroanalysis 26(6):1214–1223
97. Liu J, Oliva J, Tong K, Zhao F, Chen D, Pei Q (2017) Multi-colored light-emitting electrochemical cells based on thermal activated delayed fluorescence host. Sci Rep 7:1524
98. Mace CR, Deraney RN (2014) Manufacturing prototypes for paper-based diagnostic devices. Microfluid Nanofluid 16(5):801–809
99. Madakam S, Ramaswamy R, Tripathi S (2015) Internet of things (IOT): a literature review. J Comput Commun 3:164–173
100. Mak WC, Beni V, Turner APF (2016) Lateral-flow technology: from visual to instrumental. Trends Anal Chem 79:297–305. Past, Present and Future challenges of Biosensors and Bioanalytical tools in Analytical Chemistry: a tribute to Prof Marco Mascini
101. Mannoor MS, Tao H, Clayton JD, Sengupta A, Kaplan DL, Naik RR, Verma N, Omenetto FG, McAlpine MC (2012) Graphene-based wireless bacteria detection on tooth enamel. Nat Commun 3:763
102. Martinez AW, Phillips ST, Carrilho E, Thomas SW III, Sindi H, Whitesides GM (2008) Simple telemedicine for developing regions: camera phones and paper-based microfluidic devices for real-time, off-site diagnosis. Anal Chem 80(10):3699–3707

103. Martinez AW, Phillips ST, Nie Z, Cheng C-M, Carrilho E, Wiley BJ, Whitesides GM (2010) Programmable diagnostic devices made from paper and tape. Lab Chip 10:2499–2504
104. Matsuda Y, Shibayama S, Uete K, Yamaguchi H, Niimi T (2015) Electric conductive pattern element fabricated using commercial inkjet printer for paper-based analytical devices. Anal Chem 87(11):5762–5765. PMID: 25952643
105. Mattana G, Briand D (2016) Recent advances in printed sensors on foil. Mater Today 19(2):88–99
106. Maxwell E, Mazzeo AD, Whitesides GM (2013) Paper-based electroanalytical devices for accessible diagnostic testing. MRS Bull 38(4):309–314
107. Meier S, Tordera D, Pertegás A, Carmona C, Ortí E, Bolink H (2014) Light-emitting electrochemical cells: recent progress and future prospects. Mater Today 17:217–223
108. Meiss T, Wertschützky R, Stoeber B (2014) Rapid prototyping of resistive mems sensing devices on paper substrates. In: 2014 IEEE 27th international conference on micro electro mechanical systems (MEMS), pp 536–539
109. Meredith NA, Quinn C, Cate DM, Reilly TH, Volckens J, Henry CS (2016) Paper-based analytical devices for environmental analysis. Analyst 141:1874–1887
110. Meyer W (2013) Connectivity in point of care testing. PathCare Pathology Forum - Point of Care Vol 4 No 2, chapter 7, 1st edn. Dietrich Voight Mia Partners, Goodwood, pp 24–25
111. Mraovic M, Muck T, Pivar M, Trontelj J, Pleteršek A (2014) Humidity sensors printed on recycled paper and cardboard. Sensors 14(8):13628–13643
112. Myny K, Smout S, Rockelé M, Bhoolokam A, Ke TH, Steudel S, Cobb B, Gulati A, Rodriguez FG, Obata K, Marinkovic M, Pham D-V, Hoppe A, Gelinck GH, Genoe J, Dehaene W, Heremans P (2014) A thin-film microprocessor with inkjet print-programmable memory. Sci Rep 4:7398
113. Nassar JM, Cordero MD, Kutbee AT, Karimi MA, Sevilla GAT, Hussain AM, Shamim A, Hussain MM (2016) Paper skin multisensory platform for simultaneous environmental monitoring. Adv Mater Technol 1(1):1600004
114. Nguyen TH, Fraiwan A, Choi S. Paper-based batteries: a review (2014) Biosens Bioelectron 54:640–649
115. Nilghaz A, Guan L, Tan W, Shen W (2016) Advances of paper-based microfluidics for diagnostics–the original motivation and current status. ACS Sens 1(12):1382–1393
116. NXP Semiconductors (2017). https://www.nxp.com/products/wireless-connectivity:wireless-connectivity
117. Nyholm L, Nyström G, Mihranyan A, Strømme M (2011) Toward flexible polymer and paper-based energy storage devices. Adv Mater 23(33):3751–3769
118. Ostfeld AE, Deckman I, Gaikwad AM, Lochner CM, Arias AC (2015) Screen printed passive components for flexible power electronics. Sci Rep 5:15959
119. Pan S, Ren J, Fang X, Peng H (2016) Integration: an effective strategy to develop multifunctional energy storage devices. Adv Energy Mater 6(4):1501867
120. Phillips EA, Shen R, Zhao S, Linnes JC (2016) Thermally actuated wax valves for paper-fluidic diagnostics. Lab Chip 16:4230–4236
121. Pierre A, Deckman I, Lechene PB, Arias AC (2015) High detectivity all-printed organic photodiodes. Adv Mater 27(41):6411–6417
122. Piila T, Juusti A, Kylmänen M, Ikonen K (2015) Printed electronics and diagnostic products: printoCent designer's handbook. Neficon Finland Oy
123. PST Sensors (2017). http://www.pstsensors.com/
124. Purandare S, Gomez E, Steckl A (2014) High brightness phosphorescent organic light emitting diodes on transparent and flexible cellulose films. Nanotechnology 25:094012
125. Qi J (2017). http://technolojie.com/
126. Qu H, Hou J, Tang Y, Semenikhin O, Skorobogatiy M (2017) Thin flexible lithium-ion battery featuring graphite paper based current collectors with enhanced conductivity. Can J Chem 95(2):169–173
127. Rackus DG, Shamsi MH, Wheeler AR (2015) Electrochemistry, biosensors and microfluidics: a convergence of fields. Chem Soc Rev 44:5320–5340

128. Rahman F, Bhuiyan MZA, Ahamed SI (2016) A privacy preserving framework for {RFID} based healthcare systems. Futur Gener Comput Syst 72:339–352
129. Ren T-L, Tian H, Xie D, Yang Y (2012) Flexible graphite-on-paper piezoresistive sensors. Sensors 12(5):6685–6694
130. Sani N, Robertsson M, Cooper P, Wang X, Svensson M, Ersman PA, Norberg P, Nilsson M, Nilsson D, Liu X, Hesselbom H, Akesso L, Fahlman M, Crispin X, Engquist I, Berggren M, Gustafsson G. (2014) All-printed diode operating at 1.6 GHz. Proc Natl Acad Sci 111(33):11943–11948
131. Shafiee H, Asghar W, Inci F, Yuksekkaya M, Jahangir M, Zhang MH, Durmus NG, Gurkan UA, Kuritzkes DR, Demirci U (2015) Paper and flexible substrates as materials for biosensing platforms to detect multiple biotargets. Sci Rep 5:8719
132. Shaker G, Rida A, Safavi-Naeini S, Tentzeris MM, Nikolaou S (2011) Inkjet printing of uwb antennas on paper based substrates. In Proceedings of the 5th european conference on antennas and propagation, EUCAP 2011, pp 3001–3004
133. Sher M, Zhuang R, Demirci U, Asghar W (2017) Paper-based analytical devices for clinical diagnosis: recent advances in the fabrication techniques and sensing mechanisms. Expert Rev Mol Diagn 17:351–366
134. Shin H, Yoon B, Park IS, Kim J-M (2014) An electrothermochromic paper display based on colorimetrically reversible polydiacetylenes. Nanotechnology 25(9):094011
135. Siegel AC, Phillips ST, Wiley BJ, Whitesides GM (2009) Thin, lightweight, foldable thermochromic displays on paper. Lab Chip 9:2775–2781
136. Siegel AC, Phillips ST, Dickey MD, Lu N, Suo Z, Whitesides GM (2010) Foldable printed circuit boards on paper substrates. Adv Funct Mater 20(1):28–35
137. Silveira CM, Monteiro T, Almeida MG (2016) Biosensing with paper-based miniaturized printed electrodes - a modern trend. Biosensors 6(4):51
138. Sipilä E, Virkki J, Wang J, Sydänheimo L, Ukkonen L (2016) Brush-painting and photonic sintering of copper oxide and silver inks on wood and cardboard substrates to form antennas for uhf RFID tags. Int J Antennas Propag 2016:1–8
139. Smith S, Bezuidenhout P, Land K, Korvink JG, Mager D (2017) Printed wireless devics for low-cost, connected sensors for point-of-care applications. In: Rapid product development association of south africa (RAPDASA) 2017 conference, pp 243–249
140. Smith S, Bezuidenhout P, Mbanjwa MB, Zheng H, Conning M, Palaniyandy N, Ozoemena K, Land K (2017) Development of paper-based electrochemical sensors for water quality monitoring. In: Proceedings of SPIE, Fourth conference on sensors, MEMS, and Electro-optic systems, 100360C., vol 10036, pp 10036–10036-6
141. Sou A, Jung S, Gili E, Pecunia V, Joimel J, Fichet G, Sirringhaus H (2014) Programmable logic circuits for functional integrated smart plastic systems. Org Electron 15(11):3111–3119
142. Sritharan D, Smela E (2016) Fabrication of a miniature paper-based electroosmotic actuator. Polymers 8(11):400
143. Stauffer NW (2011). Solar cells printed on paper: http://energy.mit.edu/news/solar-cells-printed-on-paper/, mit energy initiative
144. Su W, Cook BS, Fang Y, Tentzeris M (2016) Fully inkjet-printed microfluidics: a solution to low-cost rapid three-dimensional microfluidics fabrication with numerous electrical and sensing applications. Sci Rep 6:35111
145. Tehrani Z, Korochkina T, Govindarajan S, Thomas DJ, O'Mahony J, Kettle J, Claypole TC, Gethin DT (2015) Ultra-thin flexible screen printed rechargeable polymer battery for wearable electronic applications. Org Electron 26:386–394
146. Telfer SJ, McCreary MD (2016) 42-4: invited paper: a full-color electrophoretic display. SID Symp Digest Technical Papers 47(1):574–577
147. Thinfilm (2016). http://thinfilm.no/2012/01/24/thinfilm-and-pst-sensors-to-jointly-develop-printed-temperature-tags-to-monitor-food-and-other-perishable-goods/
148. Thom NK, Yeung K, Pillion MB, Phillips ST (2012)"fluidic batteries" as low-cost sources of power in paper-based microfluidic devices. Lab Chip 12(10):1768–1770
149. Turner APF (2013) Biosensors: sense and sensibility. Chem Soc Rev 42(8):3184–3196

150. Vilmi P, Nelo M, Voutilainen J-V, Palosaari J, Pörhönen J, Tuukkanen S, Jantunen H, Juuti J, Fabritius T (2016) Fully printed memristors for a self-sustainable recorder of mechanical energy. Flex Print Electron 1(2):025002
151. VTT Technical Research Centre of Finland Ltd (2017). http://www.vttresearch.com/
152. Vyas R, Lakafosis V, Rida A, Chaisilwattana N, Travis S, Pan J, Tentzeris MM (2009) Paper-based RFID-enabled wireless platforms for sensing applications. IEEE Trans Microwave Theory Tech 57(5):1370–1382
153. Wang B, Kerr LL (2011) Dye sensitized solar cells on paper substrates. Sol Energy Mater Sol Cells 95(8):2531–2535. Photovoltaics, Solar Energy Materials & Thin Films, {IMRC} 2009-Cancun
154. Wang S, Liu N, Yang C, Liu W, Su J, Li L, Yang C, Gao Y (2015) Fully screen printed highly conductive electrodes on various flexible substrates for asymmetric supercapacitors. RSC Adv 5:85799–85805
155. Wang SQ, Chinnasamy T, Lifson MA, Inci F, Demirci U (2016) Flexible substrate-based devices for point-of-care diagnostics. Trends Biotechnol 34(11):909–921
156. Wang Y, Guo H, Chen J, Sowade E, Wang Y, Liang K, Marcus K, Baumann RR, Feng ZS (2016) Paper-based inkjet-printed flexible electronic circuits. ACS Appl Mater Interfaces 8(39):26112–26118. PMID: 27582243
157. Wang Y, Song Y, Xia Y (2016) Electrochemical capacitors: mechanism, materials, systems, characterization and applications. Chem Soc Rev 45:5925
158. Willert A, Baumann RR (2013) Customized printed batteries driving sensor applications. In: Semiconductor conference Dresden-Grenoble (ISCDG), 2013 international, pp 1–4
159. Winter M, Brodd RJ (2004) What are batteries, fuel cells, and supercapacitors? Chem Rev 104(10):4245–4270. PMID: 15669155
160. Wu H, Chiang SW, Lin W, Yang C, Li Z, Liu J, Cui X, Kang F, Wong CP (2014) Towards practical application of paper based printed circuits: capillarity effectively enhances conductivity of the thermoplastic electrically conductive adhesives. Sci Rep 4:6275
161. Xia Y, Si J, Li Z (2016) Fabrication techniques for microfluidic paper-based analytical devices and their applications for biological testing: a review. Biosens Bioelectron 77(Suppl C):774–789
162. Yamada K, Shibata H, Suzuki K, Citterio D (2017) Toward practical application of paper-based microfluidics for medical diagnostics: state-of-the-art and challenges. Lab Chip 17:1206–1249
163. Yang L, Tentzeris MM (2007) Design and characterization of novel paper-based inkjet-printed RFID and microwave structures for telecommunication and sensing applications. In: IEEE MTT-S international microwave symposium digest, pp. 1633–1636
164. Yang L, Rida A, Vyas R, Tentzeris MM (2007) RFID tag and rf structures on a paper substrate using inkjet-printing technology. IEEE Trans Microwave Theory Tech 55(12):2894–2901
165. Yang L, Zhang R, Staiculescu D, Wong CP, Tentzeris MM (2009) A novel conformal RFID-enabled module utilizing inkjet-printed antennas and carbon nanotubes for gas-detection applications. IEEE Antennas Wirel Propag Lett 8:653–656
166. Yang P-K, Lin Z-H, Pradel KC, Lin L, Li X, Wen X, He J-H, Wang ZL (2015) Paper-based origami triboelectric nanogenerators and self-powered pressure sensors. ACS Nano 9(1):901–907
167. Yao B, Yuan L, Xiao X, Zhang J, Qi Y, Zhou J, Zhou J, Hu B, Chen W (2013) Paper-based solid-state supercapacitors with pencil-drawing graphite/polyaniline networks hybrid electrodes. Nano Energy 2(6):1071–1078
168. Yao B, Zhang J, Kou T, Song Y, Liu T, Li Y (2017) Paper-based electrodes for flexible energy storage devices. Adv Sci 4(7):1700107
169. Yetisen AK, Akram MS, Lowe CR (2013) Paper-based microfluidic point-of-care diagnostic devices. Lab Chip 13:2210–2251
170. Yoon D-Y, Kim T-Y, Moon D-G (2010) Flexible top emission organic light-emitting devices using sputter-deposited ni films on copy paper substrates. Curr Appl Phys 10(4, Suppl):e135–e138

171. Yuan L, Xiao X, Ding T, Zhong J, Zhang X, Shen Y, Hu B, Huang Y, Zhou J, Wang Z (2012) Paper-based supercapacitors for self-powered nanosystems. Angew Chem Int Ed Engl 51:4934–4938
172. Yuan M, Alocilja EC, Chakrabartty S (2016) Self-powered wireless affinity-based biosensor based on integration of paper-based microfluidics and self-assembled RFID antennas. IEEE Trans Biomed Circuits Syst 10(4):799–806
173. Yun MJ, Cha SI, Seo SH, Kim HS, Lee DY (2015) Insertion of dye-sensitized solar cells in textiles using a conventional weaving process. Sci Rep 5(11022):11022
174. Zhang T, Hu M, Liu Y, Guo Q, Wang X, Zhang W, Lau W, Yang J (2016) A laser printing based approach for printed electronics. Appl Phys Lett 108(10):103501
175. Zhang J, Tian GY, Marindra AMJ, Sunny AI, Zhao AB (2017) A review of passive RFID tag antenna-based sensors and systems for structural health monitoring applications. Sensors 17(2):265
176. Zhao C, Thuo MM, Liu X (2013) A microfluidic paper-based electrochemical biosensor array for multiplexed detection of metabolic biomarkers. Sci Technol Adv Mater 14(5):054402
177. Zheng G, Cui Y, Karabulut E, Wågberg L, Zhu H, Hu L (2013) Nanostructured paper for flexible energy and electronic devices. MRS Bull 38(4):320–325
178. Zhong Q, Zhong J, Hu B, Hu Q, Zhou J, Wang ZL (2013) A paper-based nanogenerator as a power source and active sensor. Energy Environ Sci 6:1779–1784
179. Zhu H, Hu L, Cumings J, Huang J, Chen Y, Preston C, Rohrbach K (2013) Highly transparent and flexible nanopaper transistor. ACS Nano 7:2106–2113

Chapter 6
The Many Roads to an Ideal Paper-based Device

Margot Karlikow and Keith Pardee

The recent Zika and Ebola virus outbreaks highlight the need for low-cost diagnostics that can be rapidly deployed and used outside of established clinical infrastructure. This demand for robust point-of-care (POC) diagnostics is further driven by the increasing burden of drug-resistant diseases, concern for food and water safety, and bioterrorism, among many other necessities. As has been discussed in previous chapters, paper-based tests provide a simple and compelling solution to such needs. Paper-based tests and devices are themselves not new and in fact were first developed in the 1930s [1], but have only emerged recently as broadly capable tools for rapid diagnostics outside of laboratory settings.

The most widely recognized paper-based diagnostic is the home pregnancy test that came on the market in 1988 [2]. This one-step assay to assess urine for the pregnancy hormone human chorionic gonadotropin uses the chromatographic effect of paper to drive flow for the onboard immunoassay that underpins detection. While the pregnancy test has been a tremendous success, many other clinical needs require portable diagnostics that are similarly complex and sensitive and, in turn, this has given birth to a whole field dedicated to the development of paper-based, POC diagnostics [3]. Accordingly, the lateral flow assay concept has been extended to a wide range of infectious disease antigens, nucleic acid sequences, and illicit drugs from diverse sample types (e.g., sputum, blood, fecal, food).

In 2004, the Whitesides group developed the "POCKET immunoassay" [4], a microfluidic compatible immunoassay to bring enzyme-linked immunosorbent assays (ELISAs) to the field and resource-limited environments where such analytical techniques are not usually available. Three years later this was followed by another report, from the same group, that used patterned paper to create paper-based microfluidics for the simultaneous analysis of glucose and protein in urine [5].

M. Karlikow · K. Pardee (✉)
Leslie Dan Faculty of Pharmacy, University of Toronto, Toronto, ON, Canada
e-mail: keith.pardee@utoronto.ca

K. J. Land (ed.), *Paper-based Diagnostics*,
https://doi.org/10.1007/978-3-319-96870-4_6

These two papers laid the groundwork for new types of paper-based devices to meet the growing need for complex, yet portable, diagnostics. Most prominently this has included microfluidic paper-based analytical devices (μPADs) for the development of low-cost and portable devices.

Paper-based devices can be designed for diverse sensing purposes, but those for health-related POC applications have especially stringent regulatory guidelines, such as those laid out by Kettler et al. in a 2004 WHO white paper. Referred to as ASSURED standards (Affordable, Specific, Sensitive, User-friendly, Rapid and robust, Equipment-free and Deliverable) these guidelines have become a standard for the field and provide a framework for engineers and clinicians alike to develop safe and effective paper-based diagnostics [6].

The ultimate goal in the paper-based diagnostics field is the development of fully integrated devices, able to achieve sample-in answer-out processes. But in the quest for the ideal paper-based devices, many obstacles remain. Here we survey some of the most recent and innovative paper-based diagnostics and the technologies that surround them, in hopes of helping to define what would be an ideal device. Accordingly, we review the areas of ongoing tool development, such as sample handling and preparation, sensitivity, and signal amplification, to frame our discussion and share advances in the field. However, the sensing modality employed in a device in many ways defines the challenges along the developmental path. Therefore, we will often discuss tool development for paper-based devices using antibody-based detection and those using nucleic acid amplification separately. This will be followed by a discussion of key technical aspects that we believe will lead the future directions of POC diagnostics, and then look at how, as a community, the vision of low cost and accessible diagnostics can be realized.

6.1 Sample Preparation

Sample preparation is "the forgotten beginning" [7] and a major bottleneck for POC diagnostics [8]. A recent study compared 13 different commercially available or close-to-market POC devices and found that in more than half of them the sample preparation was not included [9]. Further, as late as in 2016, a review from Tang et al. highlighted the fact that most analyte extraction methods require either an external electrical power, an external pump, or multiple steps [10]. In POC settings, the processing of biological samples for diagnostics is a critical challenge complicated by the diversity and the intrinsic complexity of patient samples (e.g., blood, saliva, urine). These challenges begin with the sample collection and require the development of a workflow to separate, extract, and often amplify the analyte of interest for the paper-based diagnostic system. Each step needs to be considered as part of the whole workflow and in the context of sample, patient, and environmental factors [11]. Below we discuss each of the sample preparation steps and explore how engineering efforts are advancing sample preparation in the pursuit of the ideal paper-based device.

6.1.1 Collection of the Sample

An ideal POC diagnostic would be free of the need for skilled personnel to operate and would provide results the same day, ideally during the same hour, obviating the need for collected samples to be stored. With these target features in mind, the sample collection modes to consider should be noninvasive procedures that can be performed by untrained or only semiskilled users.

6.1.1.1 Blood

Blood is the most clinically important diagnostic sample type due to the fact that it provides a rich source of molecular information that can be monitored by either antibody or nucleic acid-based approaches [12]. By far, the most widely used technique for POC blood collection is the finger stick, which uses a blood lancet for capillary blood sampling. The technique is widely used for monitoring blood glucose in POC settings and more recently has been used to explore among others early innate transcriptomic responses generated over time after vaccination [13]. Exciting work in this space also includes the demonstration that RNA quality and gene expression data obtained using finger stick collection correlates well with more conventional venipuncture [14], which suggests a strong future for robust nucleic acid analysis at the POC.

Skin chemistry is also a "window to body's health" [15] and can be used as a substitute of blood analysis. Indeed, serum-rich interstitial fluid in the dermis contains numerous biomarkers (glucose, creatinine, lipids, nucleic acids, proteins, urea, etc.). Moreover, as a protective barrier to our bodies, the skin hosts significant immunological machinery and, as such, is an excellent potential tissue for monitoring the state or history of the immune system and infection. Skin chemistry, however, has proven difficult to monitor because of the lack of convenient methods of sampling [15].

Fortunately microneedles have recently emerged as a method to sample skin chemistry. Initially developed for transdermal drug delivery [16–18] and glucose monitoring [19], Corrie and colleagues developed an innovative POC device for the monitoring of specific biomarkers from the serum-rich fluids of the upper dermis [20]. To do so they engineered chemically and mechanically modified microprojection arrays to selectively retrieve protein biomarkers from the skin. With this minimally invasive technique for sampling soluble markers, authors demonstrated the capture of anti-FluVax® IgG in mice vaccinated with FluVax®. More recently microprojection arrays have also been developed as a wearable patch for multiplexed sampling of biomarkers directly from the skin of mice [21].

While these devices are not paper based, they are portable and have the potential to enable low-cost diagnostics in the field. Importantly microprojection arrays also have the potential to further enable pain-free sample collection and may provide a method for integrating multiplexed sample collection to established paper-based diagnostic platforms. However, important challenges remain to be solved: factors such as (1) the time required to recover samples from the skin (which can take several hours for low-abundance biomarkers [20]); (2) the design of arrays, for instance, the depth at which targeted cell types are found varies between the gender and age of patients, and accordingly the thickness of arrays/projects must be properly calibrated to avoid sampling artifacts; and (3) the analyte type to be harvested by this method. Further, reports to date have been limited to protein-based antigens. Nucleic acid sampling with microneedles/microprojections could greatly improve the development of skin chemistry as a diagnostic target. Interestingly, the recovery and analysis of nucleic acids from microprojections were described in an early patent [22].

Taken together, advances in blood and interstitial fluid collection are beginning to enable the harvest of minute samples in a way that can be done by untrained individuals in settings outside of the lab for the sampling of both proteins and nucleic acids.

6.1.1.2 Respiratory Samples

Sputum is a thick, cloudy, and sticky sample coughed up from the lower airway that contains a mixture of mucus, with endogenous and exogenous components, mucin molecules, filamentous actin, elastin fibers, bacteria, cell debris, and DNA. It can be coughed up in a natural way or be induced by inhalation of steam, sterile saline, or glycerin aerosol. Importantly, sputum samples are not saliva samples, which contain large amount of oral bacteria [23] that can contaminate sputum [24]. This complex mixture is part of the immune system that functions as a natural barrier to infection and as such is an interesting fluid to study. The value of sputum as a diagnostic fluid has even been demonstrated through the monitoring of color changes that have been correlated with lung cancer, and bacterial or viral infections. A particularly relevant use of sputum samples in diagnostics has proven to be tuberculosis. Here nucleic acid-based POC tests have been developed for the detection of tuberculosis, efficacy of treatment, and determination of coinfection with HIV [25, 26] [27, 28].

6.1.1.3 Other Sample Types

There are many other sample types that can be collected from patients, including, among others, urine, tears, stool, and buccal samples. In many cases the disease involved is an important factor in determining the best sample. Urine, for instance, is one that is easily collected and has shown significant diagnostic value. One timely example of the use of urine is as a sample for molecular diagnostics of the Zika virus.

The Zika virus has been found in blood plasma for only a median period of 10 days following infection and this has made the detection of the virus difficult for a disease where most patients are asymptomatic (Jean Michel [29]). Interestingly, Zika virus can be found in patient urine and saliva for as long as 30 days [30]. Whole blood (22 days) and semen (62 days) also appear to provide a greater diagnostic window for Zika virus than plasma [29, 31, 32].

Swab-based methods (e.g., skin and buccal) are another important means of collecting samples from patients [33]. Swabs are often used in lateral flow POC tests for respiratory infections such as influenza and streptococcal antigens (e.g., Strep test). While most tests require swabs to be eluted, shaken and pipetted on the device (like the Alere BinaxNOW RSV), a recent paper from the Yager lab reports a nasal "swab-to-result" device for diagnosis of two strains of influenza with an average time to result of 35 min [34].

As the field of POC diagnostics moves forward, there is great opportunity to probe some of these less frequently considered sample types as well as exploit pathogen physiology by selecting fluids enriched in the telltale signs of target infections. Regardless of sample type, once collected, crude samples are often too complex for direct analysis. To remove interfering agents or background, and obtain a better sample to analyze, a separation step is often required as part of sample preparation. At the POC, this can be a challenging operation and one that, while often neglected, is starting to see real progress.

6.1.2 Separation

6.1.2.1 Blood

Blood plasma, which makes up about 55% of whole blood, is extremely rich in proteins, nucleic acids, ions, and molecules, making it an invaluable target for diagnostics. However, red blood cells constitute the majority of the remaining blood volume and this complicates the separation process. If not correctly and carefully removed, red blood cells can clog, rupture, block, and foul a diagnostic device and the subsequent assay. Under laboratory conditions, blood samples are easily processed at 4 °C and centrifuged to isolate the plasma fraction without clotting. In the field or at home, the task of separating the plasma is decidedly more challenging and as a result POC blood separation techniques have fallen behind that of other sample types.

That said, there are some exciting innovative approaches that are extending the preparation of plasma from whole blood to field environments. Liu and colleagues [35] developed a plasma separator that allows the isolation of plasma from undiluted blood. This technique however requires almost 2 mL of blood to extract 300 μL of plasma and, as a result, a trained individual is needed to collect the sample. More recently a low-volume (2–4 μL) microfluidic on-chip blood separation device compatible with finger stick collection has been developed that, when compared to off-chip separation, recovers 90% of the proteins and 100% of the nucleic acids from whole blood [36].

Microdialysis has been another important method for POC blood separation. The approach uses nanofiltration, where the molecular weight cutoff of the membrane is used to selectively concentrate a wide range of targets [37–39]. However, several challenges like membrane fouling and clogging persist [40], and pressure- or voltage-driven mechanisms, among others, are still needed for devices to work. Byrnes and colleagues recently demonstrated the use of passive lateral flow and anion-exchange chromatography to simultaneously purify and concentrate DNA from samples with high protein content or blood, using porous membranes coated with chitosan [41]. While in their study membranes clogged at concentrations of below that of whole blood, they demonstrated that on-site dilution of patient samples would be a simple and easily implemented solution. Using an orthogonal approach, D'Amico and colleagues developed a microfluidic membrane-less technology to isolate and concentrate bacteria from whole blood [42] based on their dielectrophoretic behavior compared to red blood cells.

6.1.2.2 Urine (and Other Secreted Fluids)

In the case of secreted fluids like urine, tears, or sweat, there can be a need to concentrate target molecules before diagnostic analysis. A simple concentration method based on evaporation and passive pumping was developed by Walker and colleagues using microfluidics [43] where they used evaporation-induced flow to concentrate beads (pathogen proxy) and fluorescently labeled BSA. This concept was recently extended by using heat and the high surface area of paper to augment evaporation for the concentration of analytes in urine [44]. This creative approach was effective at a 20-fold concentration of a urine biomarker (TB-related glycolipid lipoarabinomannan) and it would be interesting to know if a similar approach could be used to concentrate nucleic acids. With longer biopolymers like DNA/RNA it might be possible to capture target molecules at the inflow edge, rather than at the site of evaporation, by essentially using a filtering or chromatography effect.

Another exciting approach for the concentration of diagnostic analytes has recently been demonstrated using digital microfluidics (DMF, [45]). DMF is a fluid-handling technique that uses electrical charge to move aqueous droplets around on Teflon-coated electrodes (a more detailed description of the potential for DMF is developed later in this book chapter). As a tool for analyte concentration, Choi and colleagues used DMF to perform solid-phase microextraction (SPME) of steroid hormones from urine. SPME fibers coated with C18 silica particles selectively concentrated the steroid from the complex sample matrix, allowing detection with downstream mass spectrometry analysis. While in this case DMF was not coupled to POC diagnostics, the technique does offer an interesting and potentially economical means of performing complex sample preparation in the field.

6.1.2.3 Sputum

Due to its nature, isolation of analytes from sputum samples is perhaps the most challenging of the possible matrices and yet, as we discussed, it is an important noninvasive source of diagnostic material. Laboratory-based processing of sputum samples for diagnostics involves chemical (mucolytic agents) or mechanical (sonication, vortexing) liquefaction prior to nucleic acid tests [46]. Very few portable systems have yet to be developed for mucus separation and sample extraction. One exciting example is the Cepheid GeneXpert technology that uses an automated cartridge for the ultrasonic lysis of organisms like *Mycobacterium tuberculosis* captured in a filter. This is followed by PCR-based detection of the pathogen and determination of rifampicin resistance [25, 47]. However the cost of the Cepheid Xpert instrument is not trivial, with a unit cost of at least $17,000 USD [48, 49]. A Phase I SBIR clinical trial has also recently been reported for the development of a new automated POC cassette for processing of sputum samples for nucleic acid extraction; however, the associated costs have yet to be reported (https://www.sbir.gov/sbirsearch/detail/706169). A recent study offers a potentially more accessible and minimally instrumented approach. Here sputum is disinfected and liquefied and then pathogen is mechanically lysed using a battery-operated bead beater. Using this approach DNA could be isolated for qPCR (or POC diagnostics) from spiked TB cultures in under 10 min [26].

6.1.3 Extraction

As described above, each sample type requires a tailored method for separation of the pathogen from the surrounding matrix. This effort must then be paired with an appropriate extraction step for the analyte of interest. In the lab there are many tools at our disposition to isolate the target proteins and nucleic acids from these matrices. These include mechanical (e.g., grinding, douncing, or even bead beating), thermal, or chemical methods (e.g., detergent, chelating agent, proteases). The resulting lysate is then often subjected to physical (e.g., centrifugation), chemical (e.g., phenol/chloroform), or a solid phase-based separation (e.g., nucleic acid column). Of course, these conventional methods of extraction are generally not well suited for use in the field where specialized equipment and skilled technicians may not be available. The POC diagnostics field, fortunately, has been focused on adapting many of these laboratory methods to field-appropriate techniques [50].

6.1.3.1 Heat

In a POC context, heat may be the simplest approach for nucleic acid extraction from pathogens. Boiling water is something that is almost universally available and a simple 40-s boiling water bath has been demonstrated to release cellular contents without damage to nucleic acids [51]. Similar results have been reported for the lysis of virus [52–54]. Often the thermal lysis of pathogens is sufficient and lysates can be loaded directly into the diagnostic workflow or analytical device. For example, the thermal lysis of bacterial cells in PCR buffer, followed by PCR amplification and sizing, was performed on a reusable chip in 1998 [55]. More recently, extraction and isolation of *Vibrio cholerae* from fecal samples were performed using solar-thermal energy. Here, without the need of external electrical power, a stable temperature of 95 °C was achieved in less than 5 min [56].

6.1.3.2 Enzymatic/Chemical

Enzymatic and chemical hydrolysis of capsids, cell walls, and membranes is used routinely in the lab to extract analytes from samples. Some of the most commonly used enzyme-based approaches are proteases or lysozyme, which are highly effective but do have limitations for POC applications when pathogen identity is not known. For example, in the case of lysozyme, some bacterial species (e.g., *Staphylococcus* sp.) have evolved resistance to the glycoside hydrolase in response to the enzyme's role as an antimicrobial in the innate immune system [57]. An interesting alternative without this limitation is the bacteriolytic enzyme achromopeptidase (ACP; [58, 59]). As with other proteases this enzyme can cause problems with downstream nucleic acid amplification reactions; however, ACP activity is labile to heat (>80 °C) so can be easily inactivated, allowing for subsequence molecular steps to be done without purification of nucleic acids. Importantly, ACP is stable dry-stored for easy POC distribution [60, 61].

Chemical hydrolysis of pathogens (e.g., alkaline lysis, detergents, organic solvents), while also very effective at extraction, comes with their own limitations as they must often be neutralized, removed, or diluted so as not to impede downstream analysis. Such additional complexity and potential point of failure is not an ideal scenario for POC applications [52]. Whatman FTA papers, which have lytic reagents embedded, may provide a more simplified workflow, but their use still requires extended washes to remove lytic agents before downstream molecular steps.

To address the challenge of lytic agents, Schilling and colleagues demonstrated continuous chemical lysis of bacteria in a pressure-driven microfluidic device that also performed separation of lysed versus non-lysed debris and optical detection [62]. To develop this system they used passive diffusion and formularized flow patterns and pump rates to monitor the fluorescence emitted by a fluorogenic enzyme when cells where lysed. Intact cells were sorted due to the flow rate and their inability to diffuse at the correct location in the device for detection. Similarly, Byrnes and colleagues have recently reported the development of a pressure-controlled device called SNAP (the system for nucleic acid purification) for nucleic acid extraction from whole blood mixed with a lysis buffer, at ambient temperature, in an enzyme-free manner without electricity [63]. The eluted samples yielded qPCR-compatible nucleic acids without any subsequent purification. Impressively, their device performed extraction of HIV-1 viral RNA from virions in whole blood in less than 35 min.

Another interesting alternative has been developed by the Keso lab, called filtration isolation of nucleic acid (FINA), which allows DNA extraction from whole blood lysed using the surfactant Triton X-100 (1%) in less than 2 min. Here authors carry out lysis on a Fusion five membrane (Whatman Inc.) that is backed with blotting paper to draw material through. Comprised of glass oxide fibers, the membrane has high affinity for DNA and thus serves as a capture agent to allow simple processing steps. Once lysed, the coated membrane is rinsed with NaOH to remove inhibitory proteins (e.g., hemoglobin). The residual NaOH does not inhibit downstream steps and so the DNA-bound membrane can be transferred directly to qPCR for analysis. Authors have demonstrated this approach in processing HIV-containing blood samples, where it has yielded sensitive and specific results for the detection of the virus [64–66]. Cost, however, of this approach is likely a limiting factor for global health applications, with an estimated price $50 USD/test and $3000 USD for the analyzer. Nonetheless, the technical concepts demonstrated with this and the two previous methods are compelling and hold significant potential for incorporation into lower cost paper-based devices.

6.1.3.3 Mechanical

Mechanical lysis uses force to burst cells and has been used to create some interesting options for analyte extraction. One clever design has used sharp nanoscale barbs in a microfluidic system to perform efficient lysis of cells without contaminating the sample with additional chemicals [67]. Here, using different microfluidic designs, flow and pressure of the sample can be increased or decreased without any external power. Another device, which has been commercialized, uses bead shearing/beating for rapid lysis of bacteria in small static volumes (<80 μL) or larger volumes (1 mL) in continuous-mode bacteria preparation [68]. Sold by Claremont BioSolutions, the device boasts excellent rates of lysis (>90%, 30 s) and has been used in a number of recent diagnostics-related publications [26, 69]. Each homogenizer is available for as little as $6.90 and is designed for single use [70].

Other mechanical disruption options are available, including the sonication-based approach used by GeneXpert described above for processing of sputum. The prototype for this minisonicator was first demonstrated with highly resistant spores from *Bacillus* and reported an optimal disruption for PCR analysis in only 30 s. The combined sample preparation and PCR identification were done in an impressive 15 min [71]. While these mechanical methods of disruption are stunning, they come at a cost (Claremont BioSolutions: $7; GeneXpert: $16.86) that may be a limiting factor for global health applications [48, 49] and perhaps even for domestic applications.

To address the hardware and/or consumable costs associated with mechanical disruption, there have been some clever solutions that combine microfluidics with chemistry-based extraction of nucleic acids. Baier and colleagues developed such a hand-free sample preparation device for nucleic acid analysis from cervical fluid for HPV diagnostics. The two-stage, credit card-sized device first concentrates cells and then extracts HPV-mRNA for an isothermal amplification-based detection [72]. Microfluidic origami also offers an exciting low-cost extraction option for use in remote settings. Here chemistry-based lysis is merged with a folded paper device that is operated by sequentially bringing taps together to process complex samples. The device was able to extract DNA from samples with as little as 33 bacterial CFU/mL from simulated sputum for an estimated cost of about $2 USD. Sample processing did take between 1 and 1.5 h, with or without heat, respectively, but the device is nonetheless compelling for many applications in global health [73].

6.1.3.4 Electrical

Electrical fields can also be used to induce lysis in diagnostic samples. This approach takes advantage of a well-studied phenomenon where high-strength electric fields (kV) can induce significant change in cellular membranes, leading to the formation of transient holes [74]. This effect has been used in a process called electroporation for the loading of cells with DNA or pharmacological substances [75, 76] and at higher electrical field strengths can be used to effectively kill/lyse microorganisms (99.99%; [74]). This lysis method has the advantages of no moving parts or the addition of reagents to samples. Moreover, the field value for cell lysis is dependent on the cell type (e.g., bacteria, fungi, and mammalian) and thus can be used for the selective lysis of cells.

Early attempts at the electrolysis of cells required substantial voltage in microfluidic systems (1400 V) [77]. However, with the incorporation of direct current and small, modified channel geometry voltage, the requirements for electrolysis have been reduced to a scale that can be delivered in the field by small batteries (e.g., 2–50 V) [78–80]. The potential to precisely localize electrolysis has also been exploited for diagnostics. While most lysis methods release analytes into bulk solution, which leads to dramatically lower concentrations for diagnostic assays, Besant and colleagues recently developed a device for the electrolysis of bacteria at a minimal distance to the site of detection [81]. To do this, authors built nanoscale compartments (1 nL) where cells were electrolyzed directly in next diagnostic sensors, enabling high local mRNA concentrations at the detector.

While these methods for electrolysis are currently too technical for low-cost POC deployment, they offer food for thought on what might be possible. Paper can be made conductive with the use of printable conductive inks [82], which have already been used to create paper diagnostics with electrochemical outputs [83, 84] and paper-based batteries [85]. So with a goal of looking toward an ideal paper-based device, it is exciting to think about the potential for printable all-in-one devices, with no moving parts, that could carry out both automated electrolysis and electrochemical sensing.

6.2 Overcoming Challenges in Sensitivity and Specificity

For low-cost portable diagnostics to have a meaningful impact on public health, they must incorporate high sensitivity and specificity. These requirements are highlighted by the WHO's ASSURED criteria. However, due to constraints on cost, time, size, and complexity, portable diagnostics often provide lower performance than lab-based counterparts. This fortunately is starting to change as paper-based diagnostics are beginning to extend lab-grade capabilities out into the world with detection thresholds in the low femtomolar to attomolar range [54, 69, 86].

Sensitivity refers to the threshold of detection for an analyte. For POC diagnostics an ideal device is able to identify target molecules directly from samples without concentration. Accordingly, sensitivity is consistently listed as a top priority for portable tests, and while improvements are being made a recent survey of POCs for sexually transmitted diseases highlighted that many tests lack the sensitivity needed to be useful in practical application [87]. Specificity is the measure of the rate at which devices report the correct results and is similarly of great importance because of the implications for false diagnosis. The ideal rate of specificity accordingly is 100%. These two features, sensitivity and specificity, often work against each other as is seen in the reciprocal effect observed between lateral flow speed and sensitivity [88] (Millipore). While challenging, the development of tools with high sensitivity and specificity is critical to extending the reach of lab-grade diagnostics globally and has tremendous potential for public health.

How diagnostic sensitivity and specificity can be achieved is very dependent on the mode of detection. For instance, in the case of sensitivity, antibody-based diagnostics need to accommodate the nature of protein-based analytes that are not easily replicated and thus generally rely on post-detection amplification. Conversely, nucleic acid-based diagnostic strategies generally involve amplification of target sequences and thus provide signal enhancement as part of the assay. For many applications, providing clinically relevant sensitivity in either of these modes will likely require the inclusion of simple equipment into the diagnostic workflow. Over time, we anticipate nanomaterials and molecular technologies to fulfill these current equipment-dependent roles to ultimately provide ideal devices that are truly equipment free.

6.2.1 *Antibodies: Signal Amplification*

The need for greater sensitivity in antibody-based diagnostics has been met through strategies that rely on first detecting the analyte and then the use of enzymes, metal ions, or nanoparticles to amplify the output signal from the assay reaction. Within these signal amplification modes there has been a wide range of techniques developed.

6.2.1.1 Enzymes

The first of these, enzymes, have a long history of application as sensors in diagnostics; perhaps the most widely recognized enzyme in POC diagnostics is glucose oxidase for the measurement of blood sugar [89]. However, enzymes are now also being used to amplify the signal from antibody-based detection of analytes. Examples of such enzyme-based amplification include the use of enzymes like horseradish peroxidase (HRP) [90] and alkaline phosphatase [91] to catalyze the generation of optical signal outputs from tests. An exciting new application of HPR signal enhancement was recently reported where the enzyme was used to deposit polydopamine for a high-contrast color-based signal. In HIV p24 lateral flow assays, this technique provided authors with a 1000-fold increase in signal detection [92].

The use of enzymes for signal amplification of course raises the issue of long-term stability of protein-based reagents for distribution to remote locations. In many cases, freeze-drying with cryoprotectants enables such use, but importantly recent advances have also aimed to tackle more challenging enzymes. In one report, the Yager lab demonstrated the long-term dry storage of the enzyme HRP by combining the cryoprotectant trehalose with $FeSO_4$ [93] and in a second approach, from our own lab, showed that antibody-enzyme conjugates could be manufactured on-site using a freeze-dried cell-free expression system [94].

6.2.1.2 Nanoparticles

The use of particles and nanomaterials has also proven to be effective in signal amplification for antibody-based assays. These materials include metal and silica nanoparticles [95–97], functionalized latex particles [98], magnetic particles [99, 100], quantum dots [101], and carbon nanomaterials [102], among others. Of these, gold nanoparticles (AuNPs) have a number of features that make them especially well suited to POC diagnostics, and accordingly have become widely used. AuNPs are very stable, which enables distribution and storage either in solution or in a dry powder and, despite the value of gold material itself, only extremely small amounts of the AuNPs are needed for detection. Gold nanoparticles are also easily functionalized with antibodies and other biomolecules and do not fade with exposure to light [103, 104]. Most importantly however, AuNPs have unique optical and electrical properties that make them useful as both visual and electrochemical outputs for diagnostics. The optical properties of AuNPs include a deep color that ranges from deep red to blue, depending on their diameter, that allows for diagnostic outputs to be read directly by the naked eye [103, 105]. For electrochemical assays, antigen-antibody interactions can be used to recruit AuNPs to electrodes, which augments the conductivity of electrode surfaces for diagnostic detection [106].

The most conventional use of AuNPs in diagnostics is as part of a "sandwich" immunoassay in a lateral flow device, where target antigen is first captured by a target-specific antibody and then probed by a second antibody conjugated to AuNPs. These colorful/electro-active nanoparticles serve as labels that allow users to track molecular interactions occurring at the surface of the paper device with the naked eye or companion electronics. In such lateral flow assays, positive results are usually indicated by the formation of a band or dot next to a positive control [95, 107]. Despite success of diagnostics like the pregnancy test, for many applications antibodies may lack the sensitivity for detection at clinically relevant concentrations [95].

Fortunately, the limit of detection (LOD) for these AuNP-labeled interactions can be enhanced in a number of ways. Enzymes, such as HRP, have been used to amplify the AuNP signal. In such applications, once target binding has occurred, colorimetric enzyme-substrates are added and enzyme activity yields a color change that augments LOD. This strategy has been reported to provide tenfold improvement in sensitivity [108, 109]. While helpful, limited enzyme stability and need for the construction of complex conjugates (e.g., AuNP + antibody + enzyme) could potentially pose challenges for real-world POC diagnostic applications [105].

Metal ions offer an alternative method for signal enhancement. Silver and gold ions have been used to increase the size of nanoparticles and in doing so increase the visual target for the diagnostics user. Silver enhancement takes advantage of autometallography, a process where donated electrons induce the reduction of silver ions to metallic silver [110]. In the mid-1980s, this method was used to increase the observable size of antibody-conjugated AuNPs for electron microscopy [111]. Building on this concept, silver enhancement was incorporated 5 years later in lateral flow devices where it can provide an increased LOD of 50- to 100-fold [112, 113].

6.2.2 *Nucleic Acids: Signal Amplification*

With the goal of extending the reach of lab-based clinical diagnostics to the POC, efforts toward truly portable and low-cost nucleic acid amplification tests (NAATs) are some of the most exciting. These techniques are approaching, or in some cases already achieving, 100% specificity and are increasingly available for patient care and surveillance of infectious disease in global health [100, 114]. As we will discuss, NAATs have a surprising range of diversity in their design and similarly provide variable sensitivity and specificity.

6.2.2.1 Isothermal Amplification

Polymerase chain reaction (PCR), the most common lab-based method for NAAT diagnostics, is a powerful technique that allows for the detection of minute amounts of specific nucleic acids through a series of amplification reactions that require thermal cycling. While it would be ideal to deploy PCR to POC settings for diagnostics, this task comes with significant challenges. PCR is expensive and requires specialized equipment, laboratory infrastructure, and skilled technicians, all of which are often limited at POC settings. As a result, patient samples are generally shipped for PCR tests in clinical labs, often leading to delayed reporting to patients [115].

There has been tremendous effort invested into moving this capability to the POC and much of this work has focused on the development of alternative isothermal nucleic acid amplification methods. As the name suggests, these amplification reactions operate at a single temperature, rather than thermal cycling, and as such do not require sophisticated equipment. In fact, heating for these reactions can even be provided using a chemical heater (e.g., calcium oxide and water reaction) [116]. Other benefits of isothermal methods include a simplified workflow, meaning that work can be done outside of the lab by individuals with little to no training [117]. Commercial isothermal reactions have recently become available and with these isothermal amplification of nucleic acids has begun to appear in paper-based formats [118, 119]. In this section we review isothermal amplification mechanisms and some of the ways that they have been coupled to paper-based diagnostics. For a complete review on this topic please see an excellent review by Craw and Balachandran [120].

NASBA, or nucleic acid sequence-based amplification, is an isothermal amplification developed in 1991 by Jean Compton at Cangene [121]. This non-cycling nucleic acid amplification method involves three enzymes and two primers, and works in two steps: (1) initial denaturation and primer annealing at 65 °C (2 min) followed by (2) nucleic acid amplification at 41 °C (>30 min). This method has been shown to have comparable sensitivity to RT-PCR [122] and was recently incorporated into the workflow of some of our paper-based diagnostics for the Zika virus where we found that the 65 °C step could be omitted [54].

Loop-mediated amplification assay (LAMP) was developed by Thai and colleagues in 2004 during the SARS coronavirus crisis and is one of the most commonly used isothermal amplification methods. In their seminal paper, they demonstrated high sensitivity (100%) and specificity (87%) when compared to reverse transcription PCR (RT-PCR) of viral RNA. While LAMP requires a constant operating temperature of 60–65 °C it has been developed as a POC diagnostic. This includes Liu and colleagues' work [123] who developed a single-chamber cassette for HIV detection. The cassette used a FTA membrane for isolation, purification, and concentration of the nucleic acids for amplification with RT-LAMP and detection in less than 1 h (~10 HIV particles). More recently, Rodriguez et al. extended LAMP to a paper-based method that enabled extraction of clinical nasopharyngeal samples, isothermal RNA amplification (RT-LAMP), and detection of the influenza A (H1N1) all on the same piece of paper [124].

Helicase-dependent amplification (HDA) and recombinase polymerase amplification (RPA) are two other popular isothermal amplification methods. HDA uses DNA helicases in place of high temperatures to generate the single-stranded DNA templates necessary for primer hybridization. The HDA reaction yields new dsDNA that are then also bound by helicases for the cycle to repeat [125]. RPA is an increasingly popular, easy-to-use isothermal amplification method that is both fast (5–30 min) and sensitive. Working at 37 °C–42 °C, this method uses three enzymes and a pair of primers to isothermally amplify DNA and can be extended to the detection of RNA by simply adding reverse transcriptase to the reaction. In the original paper RPA is used to detect methicillin-resistant *Staphylococcus aureus* [126] and has since been extended to the detection of a wide range of pathogens (viruses, parasites, bacteria) [127–129]. For a complete review on the use of RPA for diagnostics, please see Daher and colleagues' recent work on the subject [130].

The development of these isothermal methods represents a remarkable advancement in the development of portable and convenient field-ready nucleic acid tests. However, despite this success, there is still room for improvement of these reactions. Perhaps the most important challenge is that these methods are susceptible to contamination, which can yield off-target products and false-positive results [131]. As we will discuss below, one strategy to address this shortcoming is to add a second sequence-specific step to the detection process.

6.2.2.2 Synthetic Biology

As we have just discussed, isothermal amplification underpins most POC NAATs and, with ongoing innovation in this area, this trend can be expected to continue well into the future. However to address some of their limitations, other molecular tools are also being brought to bear in an effort to increase sensitivity, improve specificity, and add new molecular capabilities. Recent work from the field of synthetic biology has contributed to this effort. This includes work of ours, and that of colleagues, with the development of a paper-based diagnostic platform that operates using an isothermal transcription and translation reaction. The approach uses a new type of riboregulator, called a toehold switch, as a diagnostic RNA sensor. The necessary enzymes and DNA encoding the toehold switch are embedded into paper and freeze-dried to allow storage and distribution without refrigeration [132, 133]. The freeze-dried paper device is activated by rehydration and the toehold switch DNA is transcribed to form an RNA hairpin at the 5′ end of a reporter mRNA. In the presence of target RNA, the hairpin of the toehold switch becomes linear, which allows ribosomes to bind for the translation of a reporter enzyme. Here the translation of the β-galactosidase enzyme catalyzes a visible color change on the paper in as little as 20 min after rehydration. As part of the proof-of-concept work, diagnostics for the strain-specific detection of the Ebola virus (Zaire vs. Sudan) and antibiotic resistance were demonstrated.

While this new capability was exciting, it lacked the sensitivity necessary to meet real needs in the field and so efforts turned to increasing the detection threshold of the system. This led to a follow-up work in May 2016 that reported a portable molecular diagnostic for the Zika virus [54]. The similarity between Zika and Dengue viruses causes cross-reactivity in serological assays for patient antibodies and so, with the outbreak in Latin America, there was great need for a portable molecular diagnostic for the virus. To solve the challenge with sensitivity, isothermal amplification was added in the workflow upstream of the toehold switch. By combining isothermal amplification with the toehold switch, not only was there an increase in signal amplification and sensitivity, but importantly this also addressed the general problem of false positives for isothermal amplification. In this configuration the detection of isothermal amplification products is now contingent on the second sequence-dependent toehold switching event for a paper-based color change.

As with isothermal amplification methods, toehold switch design is rational and can be done rapidly using an algorithm that selects regions of pathogen's genome with low secondary structure and that are unique from closely related pathogens and the human genome. Toehold switches are able to tolerate up to 11% mismatch in base pairing, yet the above computational selection of targets ensures no homology with off-target sequences [54]. This is an important feature that accounts for the constant evolutionary drift in pathogen genomes, meaning that toehold switches designed for first sequenced American strain of the Zika virus (2016) are also, in theory, capable of detecting all strains of the virus since its discovery in 1947 while still distinguishing from closely related Dengue virus isotypes.

This molecular capability was paired with a low-cost, battery-powered electronic reader to facilitate quantification in the field. When combined, the system was able to detect Zika virus at clinically relevant concentrations (2.8 femtomolar) from viremic plasma. Further, a combination of NASBA and CRISPR/Cas9 cleavage was used to discriminate between strains of the Zika virus with single base pair resolution at the POC [54]. Having demonstrated these paper-based Zika diagnostics with live virus and infected plasma, work is continuing in a patient and field trial in Latin America to solve the challenges of deployment. Follow-on work by the Collins and Zhang labs has further extended these capabilities with new work that relies on a different Cas protein. The system called SHERLOCK for "Specific High Sensitivity Enzymatic Reporter UnLocking" was demonstrated with the identification of the Zika and Dengue viruses, pathogenic bacteria, and SNPs in the human genome [69].

Synthetic biology has also developed many cell-based sensors for applications such as heavy metal or pesticide detection [134]; however for the most part these tools have not been translated into practice because of concerns over biosafety and challenges with getting efficient transport of analytes into cells [135]. A recent paper has incorporated brewer's yeast into a paper-based dipstick that can be used with patient samples, such as blood and urine [136]. Ostrov et al. achieved this by engineering extracellular G-protein-coupled receptors on the surface of yeast to sense peptides from human, plant, and food fungal pathogens. Spotted onto paper and dried for storage, these yeast cells turn a bright orange color in response to target pathogens. While work remains to translate these sensors to practical applications, one of the most exciting aspects of this work is that this is a self-replicating diagnostic that could provide very-low-cost diagnostics in global settings.

6.3 Equipment-Free Diagnostics

One of the most debated elements of POC devices relates to its instrumentation. The ASSURED features suggested by the WHO were developed as general guidelines to assist in the development of ideal POC devices. Among those criteria, "E" indicates equipment-free devices. However, should we really exclusively focus on developing non-instrumented devices?

The call for equipment-free diagnostics is a compelling and rational goal for global applications where resources are limited (at home or in the field for example). Such designs will help to ensure robustness and low cost. Further, in POC environments with potentially limited access to electricity, it does make sense to strive for autonomous devices that are not reliant on electrical power infrastructure. This argument also extends to communities with limited or no access to smartphones with cameras. However, such scenarios are growing less, not more, common. Electric grids are becoming more reliable and in remote regions solar power, other renewable energy sources, and battery technologies are increasingly available [137, 138]. Similarly, mobile phone access is growing exponentially. There are currently over 5 billion unique mobile phone subscribers globally and this number is expected to grow by another billion in the next 5 years, and much of this growth is expected in developing regions where concern over access originally prompted the call for equipment-free diagnostics [139, 140].

Accordingly, we advocate that the equipment-free goal should be a longer term target and that providing portable, lab-grade sensitivity and specificity is most critical and should take precedence. Over time these capabilities will be progressively extended to locations with greater logistical challenges. Urbanization has brought large populations into cities where power is available [137, 141], although perhaps sporadically, and the resources are available to enable diagnostics with minimal equipment requirements. Urban populations are in great need of accessible diagnostics outside of hospitals, and we would argue that such applications can help to catalyze the long-term aspiration of complete independence from equipment. At home or in the clinician's office, this same minimal equipment-based approach for POC diagnostics would satisfy a significant unmet need in the near term and similarly drive the improvements necessary for truly equipment-free tools, where needed.

A recent paper from the Yager and Lutz labs does a great job of striking a balance between the need for equipment and the goal of equipment-free diagnostics by combining the necessary functions into a single-use cartridge [142]. Fitting within the equipment-free mandate, their system, called the multiplexable autonomous disposable nucleic acid amplification test (MAD NAAT), does not require skilled users nor rely on a smartphone for delivery or transmission of results. Other important features include a simple design where only a few manipulations are required by the user, the energy to perform the nucleic acid amplification comes from batteries within the cassette, and the device provides a simple yes/no answer. This is an exciting step toward equipment-free diagnostics and a trend that is reflected by other recent commercialization efforts.

But of course, not all clinical needs are compatible with a "yes/no" answer. Features such as quantification and test multiplexing are important milestones on the path to the ideal device [11, 143]. So while the field may be on the verge of equipment-free solutions for applications of yes/no answers, the incorporation of simple equipment will likely be important in meeting many other clinical needs at the POC. With the rise of Arduino, Raspberry Pi, and other DIY components, as well as provider-subsided mobile phones, the requirement for equipment does not necessarily mean expensive. This argument is further underscored by the complexity and presumable price of equipment-free, disposable all-in-one cartridge-type diagnostics. While such cartridge-based diagnostics may provide a promising option for unmet needs in countries where governments and health insurance cover the cost, global health applications are not likely able to bear the cost even when concessional prices are applied [49]. Thus, for now, we advocate for the research and development of diagnostics that seek a balance between high-quality results and cost through minimal equipment workflows, rather than equipment-free solutions alone.

6.3.1 Paper-Based Devices

Clever designs based on paper are serving to miniaturize some of these equipment-dependent steps into compact and inexpensive devices. Paper is a simple material and inexpensive to manufacture and its intrinsic capillary force moves liquids without the need of pump, batteries, or an external power source. Lateral flow assays, like the pregnancy test, are the most common commercial examples of where paper fulfills such a role. They provide robust and reliable performance, but are limited to a single diagnostic function and are not easily adapted to complex multistep processes. However, as we have discussed throughout the chapter, innovative researchers are extending paper materials to these more complex tasks. In an early example of this, Martinez and colleagues developed a 3D microfluidic paper analytical device (μPAD) by stacking layers of wax-patterned paper to allow for more complex molecular processes and multiplexing (fluid distribution into arrays with >1000 zones) while keeping simplicity in fabrication and cost low [144]. Wang et al. [145] developed a "pop-up" device from a folded single sheet of paper, enabling sequential control of the timing and the fluidic path.

There are also the origami devices mentioned earlier that enable complex processes like sample preparation [73]. Moreover, paper has found its way into hybrid devices where it is embedded into plastic channels as fluidic media for instrument-free use [107]. Here the addition of water and the action of folding allowed the sequential delivery of dry-stored reagents to diagnostic samples. Another creative approach has taken advantage of another cellulose-based commodity, cotton thread. Here Zhou and colleagues [146] developed an immunoassay on threads allowing multiplexing as well as quantification of the analyte. So by simply stacking, weaving, folding, or unfolding the paper/cellulose, creative 3D devices can reduce dependence for equipment by mediating sequential organization and control over sample preparation and/or detection.

6.3.2 *Paper Digital Microfluidics*

Digital microfluidics (DMF) will be a key disruptive technology for POC diagnostics on the path to minimal equipment and equipment-free workflows. Just as molecular technologies and nanomaterials promise to replace equipment by providing lower temperature diagnostic regimes and signal amplification, DMF promises to dramatically simplify low-cost fluid handling and automation of complexity. Through an array of patterned electrodes, DMF uses the application of an electrical field to dispense, move, and mix microdroplets (pico- to microliters) in a chip [147]. While most microfluidic devices rely on external pressure sources, pumps, continuous flow, and large amount of samples or reagents, DMF technology enables savings in reagents, sample, and therefore cost. In 2004, Srinivasan et al. [148] were able to actuate by electrowetting, human physiological samples like whole human blood, serum, plasma, urine, saliva, sweat, and tear as well as perform a glucose assay. The flexibility of DMF enables functions like sample separation and extraction, for example, with the use of mixing areas with lytic agents or heating zone on a single device. In 2009 Mousa et al. [149] successfully separated and extracted estradiol from breast tissue, whole blood, and serum and in 2015 Ng et al. [150] presented a proof of concept for the use of DMF as a diagnostic platform for detection of infectious diseases. The demonstration was performed with a fully integrated sample preparation for the detection of rubella infection. The shoebox-sized automated platform was able to aliquot, dilute, and multiplex four parallel assays as well as deliver an automated digital readout of the assay. Recent work has combined DMF with paper (paper-DMF) to create cheap, fast, complex, and multiplexed diagnostic devices [151] using conductive silver ink on paper.

What is especially exciting about the technology is that it is adaptive. A common electrode array can be reprogrammed after manufacturing to execute virtually any of the steps in the diagnostic workflow. All of the reagents necessary for sample preparation and analysis can be incorporated into the device, along with the preprogrammed sequence of droplet movement. Thus, the paper-DMF platform can provide sequential flow of the sample and reagents, as well as complex multistep processes, without any intervention of the user. Importantly, recent advances in the manufacture of DMF devices, with roll-to-roll production, have brought device cost down to $0.63 USD per chip [152]. Finally, a device called the DropBot has recently been released as an open-source platform for controlling droplet actuation in DMF [153]. Open-source hardware and software is an important new trend for diagnostics that allows anyone to build and modify companion devices for their own needs. So while paper-DMF is indeed instrumented, with such impressive capabilities and economics, we will likely see DMF technology embedded in many future POC diagnostics.

6.4 Future of the Technology

With the future of diagnostic technologies in mind, we asked ourselves where is paper-based device technology heading. What can we expect from the field in the next years? Based on the current literature, we propose a "what if" list of what could the technology become.

"What if" paper-based devices allowed:

- Together with the massive development of wearable electronics like watches, and even clothing (like Cityzen sciences or Hexoskin, among others), sensors that monitor vital signs are starting to be found in everyday life. Güder and colleagues recently developed a smart paper that functions as a low-cost electrical respiration sensor that is able to transmit data concerning respiration to a smartphone [154]. So "what if" smart papers were the new low-cost technology allowing routine monitoring of vital signs and other clinical features in individuals admitted to the hospital?
- Understanding and consent of the populations for research is an important challenge facing the research community. However, as paper-based diagnostic capabilities grow, an automated, anonymized electronic reporting becomes possible, and "what if" paper devices could allow large epidemiological studies through citizen science? Such studies could be health related, but they could also aid with monitoring disease vectors (e.g., ducks for influenza), environmental factors, etc. [89, 155].
- "What if" paper devices (low-cost technology) and citizen science could make a shift from responding to illness to anticipating disease with early detection and health monitoring?
- Recent work reports electrically activated paper actuators that fold and bend in response to an applied current [156]. "What if" paper-based diagnostic devices could move on their own, following a preprogrammed sequence of movements, to complete a diagnostic workflow from sample collection to the delivery of the result?
- In 2013, Safavieh and Juncker introduced the term of Capillarics, a preprogrammed, self-powered microfluidic circuit built from capillary elements where liquid handling is encoded in the structure and chemistry of the conduits [157]. "What if" paper-based devices could be similarly self-powered through careful design of their materials?
- "What if" paper-based devices, thanks to their low-cost technology and production, could permit the development of a more accessible type of personalized medicine?

- But maybe the most important point: "What if" POC diagnostic devices were designed and developed for a specific problem and in collaboration with end users? This could improve the ratio of devices capable of success at the POC. In a paper by Kumar and colleagues, two case studies exemplify the potential benefit of this approach and highlight the technical and operational lessons provided by such community-based partnerships [11].

6.5 Conclusion

Looking back over the contents of the book we are struck by the pace, scale, and diversity of ongoing work, as well as the realization that the field of POC diagnostics has grown beyond a scope that can be mastered by any one technique or investigator. This of course is exciting, but also daunting, for researchers as they seek to advance the field. The ASSURED criteria will provide ongoing guidance to the development of relevant tools, but of course it is important to recognize that each design feature should be considered in the context of the intended POC environment. In other words, the intended application of an ideal diagnostic device must be considered from the outset of design. When cost is not the main limiting factor to deployment, ultrarapid, self-contained disposable cassettes may be ideal. However in well-established but poor urban environments, other factors such as time or convenience may be less of a constraint on design in favor of being able to provide clinical-grade results. There may even be a future for applications where even sensitivity or specificity can be sacrificed in favor of strategies in which multiplexed detection and low cost provide redundancy for diagnosis of complex disease (e.g., paper-based gene expression profiling; [158]). The key is that without a final set of applications and target population in mind, the complete design space available to build practical portable diagnostics may not be recognized (Fig. 6.1).

Similarly, as developers of POC diagnostics we must also be held to account on implementation, including a workflow from sample through to answer. Steps toward implementation must follow technology development; otherwise our efforts become only an academic exercise, which must not happen. Unfortunately, the literature is replete with reports of incremental advances that have not been demonstrated with real samples and/or under real-field conditions. While the reasons for this are complex, future leaders in the field, and those who will see their designs catch the attention of the funders, will be the ones who take the time (and risk) to push the limits of their technology and iterate until their design criteria are met. This may require a series of studies and vertical integration of our laboratory efforts, but the responsibility is on us to demonstrate the value of our work at point of care.

Other trends that will continue to be important to the advancement of our field will be the adoption and adaptation of disruptive technologies from other areas of study. The field is built upon this approach and as we seek to extend our tools to new populations, environments, and clinical needs, we will need to continue to combine these innovations into our designs. Technologies that we see as key areas

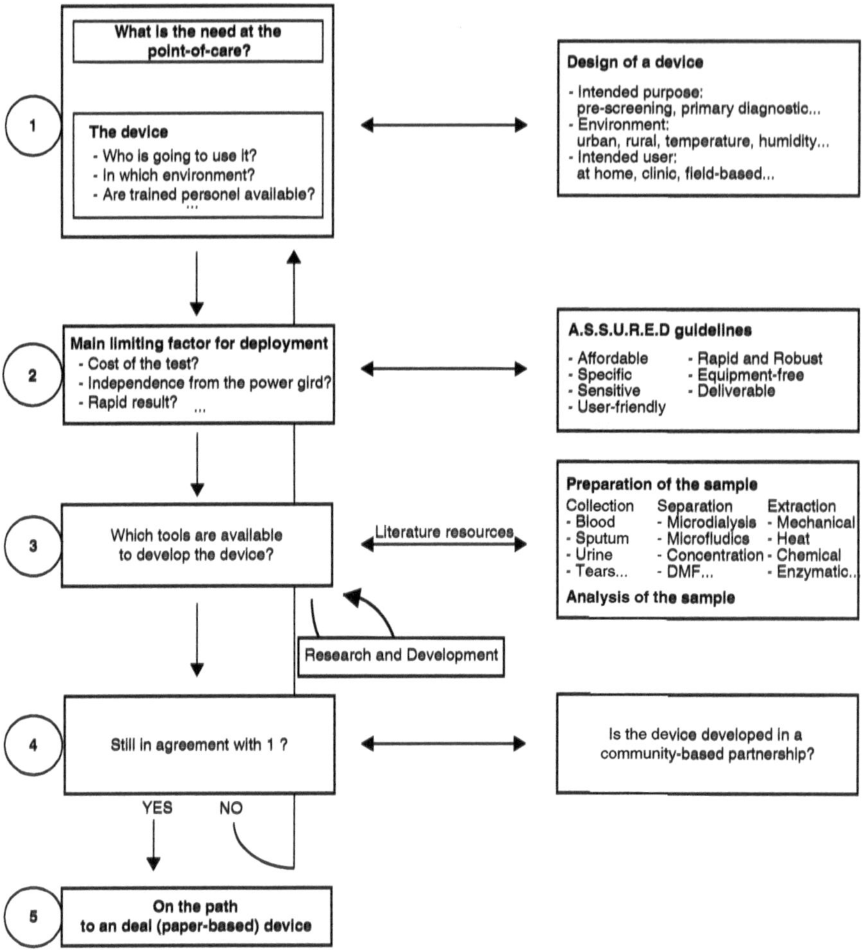

Fig. 6.1 Design of an ideal (paper-based) device workflow. The successful design and implementation of paper-based diagnostics require a holistic view that considers the goals of the test, the limitations in the intended environment for its use, and the context of the sample from which it will be expected to perform. This technical design and development must also be paired with partnerships at the community level and, ultimately, tested under field conditions

to watch are the following. (1) Sample preparation technologies that are portable and cost effective are perhaps the most critical to the success of our field as well as the compatibility of the preparation, extraction, and analysis steps. It is this capability that will convert innovative assays into field-ready tools. Here we see digital/microfluidics as key to implementing low-cost and automated workflows that can be used by nonexperts under field conditions. (2) Nucleic acid amplification and other molecular technologies will be an increasing source of capability for POC diagnostics as researchers incorporate new techniques and enzymes, as well as methods for deployment without refrigeration. (3) Materials engineering and nanotechnologies will also provide exciting and increasingly cost-effective ways of moving laboratory assays to the field.

Taken together, the merging of new sensor technologies with the incredible advances in materials, electronics, and telecommunications suggests that the field of POC diagnostics is on the verge of a revolution. This potential boon in capabilities could not come at a better time. The human population is growing at a rate of 1.6 million every week, and so, with such growth, our ability to provide timely access to health care will increasingly depend on distributed care through portable tools of diagnosis. Moreover, this drive to reduce cost and extend access to health care is something that will be globally beneficial and our challenge as a field will be how to meet such diverse needs. And it is through this lens that we suggest that the journey to the ideal device(s) has many paths.

References

1. Yagoda H (1937) Applications of confined spot tests in analytical chemistry: preliminary paper. Ind Eng Chem Anal Ed 9:79–82
2. Davies RJ, Eapen SS, Carlisle SJ (2008) Lateral-flow Immunochromatographic assays. In: Handbook of biosensors and biochips. John Wiley & Sons, Ltd, New Jersey
3. Anderson CE, Shah KG, Yager P (2017) Sensitive protein detection and quantification in paper-based microfluidics for the point of care. Methods Enzymol 589:383–411
4. Sia SK, Linder V, Parviz BA, Siegel A, Whitesides GM (2004) An integrated approach to a portable and low-cost immunoassay for resource-poor settings. Angew Chem Int Ed 43:498–502
5. Martinez AW, Phillips ST, Butte MJ, Whitesides GM (2007) Patterned paper as a platform for inexpensive, low-volume, portable bioassays. Angew Chem Int Ed 46:1318–1320
6. WHO (2004) TDR|Mapping the landscape of diagnostics for sexually transmitted infections [WWW Document]. WHO. http://www.who.int/tdr/publications/tdr-research-publications/mapping-landscape-sti/en/. Accessed 7 July 2017
7. Brehm-Stecher B, Young C, Jaykus L-A, Tortorello ML (2009) Sample preparation: the forgotten beginning. J Food Prot 72:1774–1789
8. Mariella R (2008) Sample preparation: the weak link in microfluidics-based biodetection. Biomed Microdevices 10:777–784
9. Niemz A, Ferguson TM, Boyle DS (2011) Point-of-care nucleic acid testing for infectious diseases. Trends Biotechnol 29:240–250
10. Tang RH, Yang H, Choi JR, Gong Y, Feng SS, Pingguan-Murphy B, Huang QS, Shi JL, Mei QB, Xu F (2016) Advances in paper-based sample pretreatment for point-of-care testing. Crit Rev Biotechnol 1–18
11. Kumar AA, Hennek JW, Smith BS, Kumar S, Beattie P, Jain S, Rolland JP, Stossel TP, Chunda-Liyoka C, Whitesides GM (2015) From the bench to the field in low-cost diagnostics: two case studies. Angew Chem Int Ed Engl 54:5836–5853
12. Anderson NL, Anderson NG (2002) The human plasma proteome: history, character, and diagnostic prospects. Mol Cell Proteomics 1:845–867
13. Obermoser G, Presnell S, Domico K, Xu H, Wang Y, Anguiano E, Thompson-Snipes L, Ranganathan R, Zeitner B, Bjork A, Anderson D, Speake C, Ruchaud E, Skinner J, Alsina L, Sharma M, Dutartre H, Cepika A, Israelsson E, Nguyen P, Nguyen Q-A, Harrod AC, Zurawski SM, Pascual V, Ueno H, Nepom GT, Quinn C, Blankenship D, Palucka K, Banchereau J,

Chaussabel D (2013) Systems scale interactive exploration reveals quantitative and qualitative differences in response to influenza and pneumococcal vaccines. Immunity 38:831–844
14. Robison EH, Mondala TS, Williams AR, Head SR, Salomon DR, Kurian SM (2009) Whole genome transcript profiling from fingerstick blood samples: a comparison and feasibility study. BMC Genomics 10:617
15. Paliwal S, Hwang BH, Tsai KY, Mitragotri S (2013) Diagnostic opportunities based on skin biomarkers. Eur J Pharm Sci 50:546–556
16. Fernando GJP, Chen X, Prow TW, Crichton ML, Fairmaid EJ, Roberts MS, Frazer IH, Brown LE, Kendall MAF (2010) Potent immunity to low doses of influenza vaccine by probabilistic guided micro-targeted skin delivery in a mouse model. PLoS One 5:e10266
17. Matriano JA, Cormier M, Johnson J, Young WA, Buttery M, Nyam K, Daddona PE (2002) Macroflux microprojection array patch technology: a new and efficient approach for intracutaneous immunization. Pharm Res 19:63–70
18. Prausnitz MR (2004) Microneedles for transdermal drug delivery. Adv Drug Deliv Rev 56:581–587
19. Wang PM, Cornwell M, Prausnitz MR (2005) Minimally invasive extraction of dermal interstitial fluid for glucose monitoring using microneedles. Diabetes Technol Ther 7:131–141
20. Corrie SR, Fernando GJP, Crichton ML, Brunck MEG, Anderson CD, Kendall MAF (2010) Surface-modified microprojection arrays for intradermal biomarker capture, with low non-specific protein binding. Lab Chip 10:2655–2658
21. Lee KT, Muller DA, Coffey JW, Robinson KJ, McCarthy JS, Kendall MAF, Corrie SR (2014) Capture of the circulating Plasmodium falciparum biomarker HRP2 in a multiplexed format, via a wearable skin patch. Anal Chem 86:10474–10483
22. Matriano JA, Cormier MJN (2004). Method for transdermal nucleic acid sampling. US6749575 B2
23. Aas JA, Paster BJ, Stokes LN, Olsen I, Dewhirst FE (2005) Defining the normal bacterial Flora of the oral cavity. J Clin Microbiol 43:5721–5732
24. Muzanye G, Morgan K, Johnson J, Mayanja-Kizza H (2009) Impact of mouth rinsing before sputum collection on culture contamination. Afr Health Sci 9:200
25. Boehme CC, Nabeta P, Hillemann D, Nicol MP, Shenai S, Krapp F, Allen J, Tahirli R, Blakemore R, Rustomjee R, Milovic A, Jones M, O'Brien SM, Persing DH, Ruesch-Gerdes S, Gotuzzo E, Rodrigues C, Alland D, Perkins MD (2010a) Rapid molecular detection of tuberculosis and rifampin resistance. N Engl J Med 363:1005–1015
26. Ferguson TM, Weigel KM, Lakey Becker A, Ontengco D, Narita M, Tolstorukov I, Doebler R, Cangelosi GA, Niemz A (2016a) Pilot study of a rapid and minimally instrumented sputum sample preparation method for molecular diagnosis of tuberculosis. Sci Rep 6:19541
27. Chanda-Kapata P, Kapata N, Klinkenberg E, Grobusch MP, Cobelens F (2017) The prevalence of HIV among adults with pulmonary TB at a population level in Zambia. BMC Infect Dis 17:236
28. Perkins MD, Cunningham J (2007) Facing the crisis: improving the diagnosis of tuberculosis in the HIV era. J Infect Dis 196:S15–S27
29. Mansuy JM, Mengelle C, Pasquier C, Chapuy-Regaud S, Delobel P, Martin-Blondel G, Izopet J (2017) Zika virus infection and prolonged Viremia in whole-blood specimens. Emerg Infect Dis 23:863–865
30. Barzon L, Pacenti M, Berto A, Sinigaglia A, Franchin E, Lavezzo E, Brugnaro P, Palù G (2016) Isolation of infectious Zika virus from saliva and prolonged viral RNA shedding in a traveller returning from the Dominican Republic to Italy, January 2016. Eur Secur 21:30159
31. Atkinson B, Hearn P, Afrough B, Lumley S, Carter D, Aarons EJ, Simpson AJ, Brooks TJ, Hewson R (2016) Detection of Zika virus in semen. Emerg Infect Dis 22:940
32. Mansuy JM, Dutertre M, Mengelle C, Fourcade C, Marchou B, Delobel P, Izopet J, Martin-Blondel G (2016) Zika virus: high infectious viral load in semen, a new sexually transmitted pathogen? Lancet Infect Dis 16:405

33. Panpradist N, Toley BJ, Zhang X, Byrnes S, Buser JR, Englund JA, Lutz BR (2014) Swab sample transfer for point-of-care diagnostics: characterization of swab types and manual agitation methods. PLoS One 9:e105786
34. Huang S, Abe K, Bennett S, Liang T, Ladd PD, Yokobe L, Anderson CE, Shah K, Bishop J, Purfield M, Kauffman PC, Paul S, Welch AE, Strelitz B, Follmer K, Pullar K, Sanchez-Erebia L, Gerth-Guyette E, Domingo GJ, Klein E, Englund JA, Fu E, Yager P (2017) Disposable autonomous device for rapid swab-to-result diagnosis of influenza. Anal Chem
35. Liu C, Mauk M, Gross R, Bushman FD, Edelstein PH, Collman RG, Bau HH (2013) Membrane-based, sedimentation-assisted plasma separator for point-of-care applications. Anal Chem 85:10463–10470
36. Son JH, Lee SH, Hong S, Park S, Lee J, Dickey AM, Lee LP (2014) Hemolysis-free blood plasma separation. Lab Chip 14:2287–2292
37. Chatterjee A, Mirer PL, Zaldivar Santamaria E, Klapperich C, Sharon A, Sauer-Budge AF (2010) RNA isolation from mammalian cells using porous polymer monoliths: an approach for high-throughput automation. Anal Chem 82:4344–4356
38. Song S, Singh AK, Kirby BJ (2004) Electrophoretic concentration of proteins at laser-patterned nanoporous membranes in microchips. Anal Chem 76:4589–4592
39. Zhu L, Zhang Q, Feng H, Ang S, Chau FS, Liu W-T (2004) Filter-based microfluidic device as a platform for immunofluorescent assay of microbial cells. Lab Chip 4:337–341
40. Van der Bruggen B, Mänttäri M, Nyström M (2008) Drawbacks of applying nanofiltration and how to avoid them: a review. Sep Purif Technol 63:251–263
41. Byrnes SA, Bishop JD, Lafleur L, Buser JR, Lutz B, Yager P (2015) One-step purification and concentration of DNA in porous membranes for point-of-care applications. Lab Chip 15:2647–2659
42. D'Amico L, Ajami NJ, Adachi JA, Gascoyne PRC, Petrosino JF (2017) Isolation and concentration of bacteria from blood using microfluidic membraneless dialysis and dielectrophoresis. Lab Chip 17:1340–1348
43. Walker GM, Beebe DJ (2002) An evaporation-based microfluidic sample concentration method. Lab Chip 2:57–61
44. Wong SY, Cabodi M, Rolland J, Klapperich CM (2014) Evaporative concentration on a paper-based device to concentrate analytes in a biological fluid. Anal Chem 86:11981–11985
45. Choi K, Boyacı E, Kim J, Seale B, Barrera-Arbelaez L, Pawliszyn J, Wheeler AR (2016) A digital microfluidic interface between solid-phase microextraction and liquid chromatography-mass spectrometry. J Chromatogr A 1444:1–7
46. Allen V, Nicol MP, Tow LA (2016) Sputum processing prior to *Mycobacterium tuberculosis* detection by culture or nucleic acid amplification testing: a narrative review. Res Rev J Microbiol Biotechnol 17(Suppl 1):69
47. Boehme CC, Nicol MP, Nabeta P, Michael JS, Gotuzzo E, Tahirli R, Gler MT, Blakemore R, Worodria W, Gray C, Huang L, Caceres T, Mehdiyev R, Raymond L, Whitelaw A, Sagadevan K, Alexander H, Albert H, Cobelens F, Cox H, Alland D, Perkins MD (2011) Feasibility, diagnostic accuracy, and effectiveness of decentralised use of the Xpert MTB/RIF test for diagnosis of tuberculosis and multidrug resistance: a multicentre implementation study. Lancet 377:1495–1505
48. Mirzayev F (2012) Current dynamics in the Xpert MTB/RIF assay pricing mechanisms [WWW Document]. http://www.stoptb.org/wg/gli/assets/html/day%202/Mirzayev%20-%20Xpert%20cartridge%20price%20dynamics.pdf. Accessed 27 June 17
49. Puri L, Oghor C, Denkinger CM, Pai M (2016) Xpert MTB/RIF for tuberculosis testing: access and price in highly privatised health markets. Lancet Glob Health 4:e94–e95
50. Kim J, Johnson M, Hill P, Gale BK (2009) Microfluidic sample preparation: cell lysis and nucleic acid purification. Integr Biol (Camb) 1:574–586
51. Sambrook J, Russel DW (2001) Molecular cloning a LABORATORY MANUAL. Cold Spring Harbor, New York
52. Heiniger EK, Buser JR, Mireles L, Zhang X, Ladd PD, Lutz BR, Yager P (2016) Comparison of point-of-care-compatible lysis methods for bacteria and viruses. J Microbiol Methods 128:80–87

53. Lee W-C, Lien K-Y, Lee G-B, Lei H-Y (2008) An integrated microfluidic system using magnetic beads for virus detection. Diagn Microbiol Infect Dis 60:51–58
54. Pardee K, Green AA, Takahashi MK, Braff D, Lambert G, Lee JW, Ferrante T, Ma D, Donghia N, Fan M, Daringer NM, Bosch I, Dudley DM, O'Connor DH, Gehrke L, Collins JJ (2016a) Rapid, low-cost detection of zika virus using programmable biomolecular components. Cell 165:1255–1266
55. Waters LC, Jacobson SC, Kroutchinina N, Khandurina J, Foote RS, Ramsey JM (1998) Microchip device for cell lysis, multiplex PCR amplification, and electrophoretic sizing. Anal Chem 70:158–162
56. Gumus A, Ahsan S, Dogan B, Jiang L, Snodgrass R, Gardner A, Lu Z, Simpson K, Erickson D (2016) Solar-thermal complex sample processing for nucleic acid based diagnostics in limited resource settings. Biomed Opt Express 7:1974–1984
57. Bera A, Herbert S, Jakob A, Vollmer W, Götz F (2005) Why are pathogenic staphylococci so lysozyme resistant? The peptidoglycan O-acetyltransferase OatA is the major determinant for lysozyme resistance of *Staphylococcus aureus*. Mol Microbiol 55:778–787
58. Leonard RB, Carroll KC (1997) Rapid lysis of gram-positive cocci for pulsed-field gel electrophoresis using achromopeptidase. Diagn Mol Pathol 6:288–291
59. Slifkin M, Cumbie R (1987) Serogrouping single colonies of beta-hemolytic streptococci with achromopeptidase extraction. [WWW Document]. http://jcm.asm.org/content/25/8/1555.short. Accessed 7 July 2017
60. Buser JR, Zhang X, Byrnes SA, Ladd PD, Heiniger EK, Wheeler MD, Bishop JD, Englund JA, Lutz B, Weigl BH, Yager P (2016) A disposable chemical heater and dry enzyme preparation for lysis and extraction of DNA and RNA from microorganisms. Anal Methods 8:2880–2886
61. Hilligoss D, Keller LM, Ramadan S, Coady J, Hellyer TJ (2011) Use of achromopeptidase for lysis at room temperature. WO2011115975 A2
62. Schilling EA, Kamholz AE, Yager P (2002) Cell Lysis and protein extraction in a microfluidic device with detection by a Fluorogenic enzyme assay. Anal Chem 74:1798–1804
63. Byrnes S, Fan A, Trueb J, Jareczek F, Mazzochette M, Sharon A, Sauer-Budge AF, Klapperich CM (2013) A portable, pressure driven, room temperature nucleic acid extraction and storage system for point of care molecular diagnostics. Anal Methods 5:3177–3184
64. Jangam SR, Agarwal AK, Sur K, Kelso DM (2013) A point-of-care PCR test for HIV-1 detection in resource-limited settings. Biosens Bioelectron 42:69–75
65. Jangam SR, Yamada DH, McFall SM, Kelso DM (2009) Rapid, point-of-care extraction of human immunodeficiency virus type 1 proviral DNA from whole blood for detection by real-time PCR. J Clin Microbiol 47:2363–2368
66. McFall SM, Wagner RL, Jangam SR, Yamada DH, Hardie D, Kelso DM (2015) A simple and rapid DNA extraction method from whole blood for highly sensitive detection and quantitation of HIV-1 proviral DNA by real-time PCR. J Virol Methods 214:37–42
67. Di Carlo D, Jeong K-H, Lee LP (2003) Reagentless mechanical cell lysis by nanoscale barbs in microchannels for sample preparation. Lab Chip 3:287–291
68. Doebler RW, Erwin B, Hickerson A, Irvine B, Woyski D, Nadim A, Sterling JD (2009) Continuous-flow, rapid Lysis devices for biodefense nucleic acid diagnostic systems. J Assoc Lab Autom 14:119–125
69. Gootenberg JS, Abudayyeh OO, Lee JW, Essletzbichler P, Dy AJ, Joung J, Verdine V, Donghia N, Daringer NM, Freije CA, Myhrvold C, Bhattacharyya RP, Livny J, Regev A, Koonin EV, Hung DT, Sabeti PC, Collins JJ, Zhang F (2017) Nucleic acid detection with CRISPR-Cas13a/C2c2. Science 356:438–442
70. ClaremontBio. OmniLyse® Lysis Cartridges|ClaremontBio.com [WWW Document]. URL http://www.claremontbio.com/OmniLyse_Cell_Disruption_Kits_s/27.htm. Accessed 7 July 2017
71. Belgrader P, Hansford D, Kovacs GT, Venkateswaran K, Mariella R, Milanovich F, Nasarabadi S, Okuzumi M, Pourahmadi F, Northrup MA (1999) A minisonicator to rapidly disrupt bacterial spores for DNA analysis. Anal Chem 71:4232–4236

72. Baier T, Hansen-Hagge TE, Gransee R, Crombé A, Schmahl S, Paulus C, Drese KS, Keegan H, Martin C, O'Leary JJ, Furuberg L, Solli L, Grønn P, Falang IM, Karlgård A, Gulliksen A, Karlsen F (2009) Hands-free sample preparation platform for nucleic acid analysis. Lab Chip 9:3399–3405
73. Govindarajan AV, Ramachandran S, Vigil GD, Yager P, Böhringer KF (2012) A low cost point-of-care viscous sample preparation device for molecular diagnosis in the developing world; an example of microfluidic origami. Lab Chip 12:174–181
74. Hülsheger H, Potel J, Niemann EG (1983) Electric field effects on bacteria and yeast cells. Radiat Environ Biophys 22:149–162
75. Cadossi R, Ronchetti M, Cadossi M (2014) Locally enhanced chemotherapy by electroporation: clinical experiences and perspective of use of electrochemotherapy. Future Oncol Lond Engl 10:877–890
76. Luft C, Ketteler R (2015) Electroporation knows no boundaries: the use of electrostimulation for siRNA delivery in cells and tissues. J Biomol Screen 20:932–942
77. Gao J, Yin X-F, Fang Z-L (2004) Integration of single cell injection, cell lysis, separation and detection of intracellular constituents on a microfluidic chip. Lab Chip 4:47–52
78. Ameri SK, Singh PK, Dokmeci MR, Khademhosseini A, Xu Q, Sonkusale SR (2014) All electronic approach for high-throughput cell trapping and lysis with electrical impedance monitoring. Biosens Bioelectron 54:462–467
79. Lee DW, Cho YH (2007) A continuous electrical cell lysis device using a low dc voltage for a cell transport and rupture - ScienceDirect [WWW Document]. URL http://www.sciencedirect.com/science/article/pii/S0925400506008033. Accessed 7 July 2017
80. Wang H-Y, Banada PP, Bhunia AK, Lu C (2007) Rapid electrical lysis of bacterial cells in a microfluidic device. Methods Mol Biol 385:23–35
81. Besant JD, Das J, Sargent EH, Kelley SO (2013) Proximal bacterial lysis and detection in nanoliter wells using electrochemistry. ACS Nano 7:8183–8189
82. Lessing J, Glavan AC, Walker SB, Keplinger C, Lewis JA, Whitesides GM (2014) Inkjet printing of conductive inks with high lateral resolution on omniphobic "R(F) paper" for paper-based electronics and MEMS. Adv Mater 26:4677–4682
83. Glavan AC, Christodouleas DC, Mosadegh B, Yu HD, Smith BS, Lessing J, Fernández-Abedul MT, Whitesides GM (2014) Folding analytical devices for electrochemical ELISA in hydrophobic R(H) paper. Anal Chem 86:11999–12007
84. Nie Z, Nijhuis CA, Gong J, Chen X, Kumachev A, Martinez AW, Narovlyansky M, Whitesides GM (2010) Electrochemical sensing in paper-based microfluidic devices. Lab Chip 10:477–483
85. Nguyen TH, Fraiwan A, Choi S (2014) Paper-based batteries: a review. Biosens Bioelectron 54:640–649
86. Cunningham JC, DeGregory PR, Crooks RM (2016) New functionalities for paper-based sensors lead to simplified user operation, lower limits of detection, and new applications. Annu Rev Anal Chem (Palo Alto, Calif) 9:183–202
87. Gaydos C, Hardick J (2014) Point of care diagnostics for sexually transmitted infections: perspectives and advances. Expert Rev Anti-Infect Ther 12:657–672
88. Hu J, Wang S, Wang L, Li F, Pingguan-Murphy B, Lu TJ, Xu F (2014) Advances in paper-based point-of-care diagnostics. Biosens Bioelectron 54:585–597
89. Cate DM, Adkins JA, Mettakoonpitak J, Henry CS (2015) Recent developments in paper-based microfluidic devices. Anal Chem 87:19–41
90. Parolo C, de la Escosura-Muñiz A, Merkoçi A (2013a) Enhanced lateral flow immunoassay using gold nanoparticles loaded with enzymes. Biosens Bioelectron 40:412–416
91. Gerbers R, Foellscher W, Chen H, Anagnostopoulos C, Faghri M (2014) A new paper-based platform technology for point-of-care diagnostics. Lab Chip 14:4042–4049
92. Li J, Baird MA, Davis MA, Tai W, Zweifel LS, Waldorf KMA, Gale M Jr, Rajagopal L, Pierce RH, Gao X (2017) Dramatic enhancement of the detection limits of bioassays via ultrafast deposition of polydopamine. Nat Biomed Eng 1:0082
93. Ramachandran S, Fu E, Lutz B, Yager P (2014) Long-term dry storage of an enzyme-based reagent system for ELISA in point-of-care devices. Analyst 139:1456–1462

94. Pardee K, Slomovic S, Nguyen PQ, Lee JW, Donghia N, Burrill D, Ferrante T, McSorley FR, Furuta Y, Vernet A, Lewandowski M, Boddy CN, Joshi NS, Collins JJ (2016b) Portable, on-demand biomolecular manufacturing. Cell 167:248–259.e12
95. Fu E, Liang T, Houghtaling J, Ramachandran S, Ramsey SA, Lutz B, Yager P (2011) Enhanced sensitivity of lateral flow tests using a two-dimensional paper network format. Anal Chem 83:7941–7946
96. Rohrman BA, Leautaud V, Molyneux E, Richards-Kortum RR (2012) A lateral flow assay for quantitative detection of amplified HIV-1 RNA. PLoS One 7:e45611
97. Yuan L, Hua X, Wu Y, Pan X, Liu S (2011) Polymer-functionalized silica nanosphere labels for ultrasensitive detection of tumor necrosis factor-alpha. Anal Chem 83:6800–6809
98. Nielsen K, Yu WL, Lin M, Davis SAN, Elmgren C, Mackenzie R, Tanha J, Li S, Dubuc G, Brown EG, Keleta L, Pasick J (2007) Prototype single step lateral flow technology for detection of avian influenza virus and chicken antibody to avian influenza virus. J Immunoassay Immunochem 28:307–318
99. Connolly R, O'Kennedy R (2017) Magnetic lateral flow immunoassay test strip development — considerations for proof of concept evaluation. Methods San Diego Calif 116:132–140
100. Zhao Y, Chen F, Li Q, Wang L, Fan C (2015) Isothermal amplification of nucleic acids. Chem Rev 115:12491–12545
101. Chun P (2009) Colloidal gold and other labels for lateral flow immunoassays. In: Lateral flow immunoassay. Humana Press, New York
102. Du D, Wang L, Shao Y, Wang J, Engelhard MH, Lin Y (2011) Functionalized graphene oxide as a nanocarrier in a multienzyme labeling amplification strategy for ultrasensitive electrochemical immunoassay of phosphorylated p53 (S392). Anal Chem 83:746–752
103. Koczula KM, Gallotta A (2016) Lateral flow assays. Essays Biochem 60:111–120
104. Wilson R (2008) The use of gold nanoparticles in diagnostics and detection. Chem Soc Rev 37:2028–2045
105. Quesada-González D, Merkoçi A (2015) Nanoparticle-based lateral flow biosensors. Biosens Bioelectron 73:47–63
106. Ge X, Asiri AM, Du D, Wen W, Wang S, Lin Y (2014) Nanomaterial-enhanced paper-based biosensors. TrAC, Trends in Anal Chem 58:31–39
107. Fu E, Liang T, Spicar-Mihalic P, Houghtaling J, Ramachandran S, Yager P (2012a) Two-dimensional paper network format that enables simple multistep assays for use in low-resource settings in the context of malaria antigen detection. Anal Chem 84:4574–4579
108. Parolo C, Medina-Sánchez M, de la Escosura-Muñiz A, Merkoçi A (2013b) Simple paper architecture modifications lead to enhanced sensitivity in nanoparticle based lateral flow immunoassays. Lab Chip 13:386–390
109. Zhang C, Zhang Y, Wang S (2006) Development of multianalyte flow-through and lateral-flow assays using gold particles and horseradish peroxidase as tracers for the rapid determination of carbaryl and endosulfan in agricultural products. J Agric Food Chem 54:2502–2507
110. Danscher G, Nørgaard JO, Baatrup E (1987) Autometallography: tissue metals demonstrated by a silver enhancement kit. Histochemistry 86:465–469
111. Scopsi L, Larsson LI, Bastholm L, Nielsen MH (1986) Silver-enhanced colloidal gold probes as markers for scanning electron microscopy. Histochemistry 86:35–41
112. Cho I-H, Seo S-M, Paek E-H, Paek S-H (2010) Immunogold-silver staining-on-a-chip biosensor based on cross-flow chromatography. J Chromatogr B Analyt Technol Biomed Life Sci 878:271–277
113. Horton JK, Swinburne S, O'Sullivan MJ (1991) A novel, rapid, single-step immunochromatographic procedure for the detection of mouse immunoglobulin. J Immunol Methods 140:131–134
114. Smit PW, Elliott I, Peeling RW, Mabey D, Newton PN (2014) An overview of the clinical use of filter paper in the diagnosis of tropical diseases. Am J Trop Med Hyg 90:195–210
115. Ghani AC, Burgess DH, Reynolds A, Rousseau C (2015) Expanding the role of diagnostic and prognostic tools for infectious diseases in resource-poor settings. Nature 528(7580):S50-S502

116. Curtis KA, Rudolph DL, Nejad I, Singleton J, Beddoe A, Weigl B, LaBarre P, Owen SM (2012) Isothermal amplification using a chemical heating device for point-of-care detection of HIV-1. PLoS One 7:e31432
117. Yan L, Zhou J, Zheng Y, Gamson AS, Roembke BT, Nakayama S, Sintim HO (2014) Isothermal amplified detection of DNA and RNA. Mol BioSyst 10:970–1003
118. Gan W, Zhuang B, Zhang P, Han J, Li C-X, Liu P (2014) A filter paper-based microdevice for low-cost, rapid, and automated DNA extraction and amplification from diverse sample types. Lab Chip 14:3719–3728
119. Linnes JC, Fan A, Rodriguez NM, Lemieux B, Kong H, Klapperich CM (2014) Paper-based molecular diagnostic for chlamydia trachomatis. RSC Adv 4:42245–42251
120. Craw P, Balachandran W (2012) Isothermal nucleic acid amplification technologies for point-of-care diagnostics: a critical review. Lab Chip 12:2469–2486
121. Compton J (1991) Nucleic acid sequence-based amplification. Nature 350:91–92
122. Houde A, Leblanc D, Poitras E, Ward P, Brassard J, Simard C, Trottier Y-L (2006) Comparative evaluation of RT-PCR, nucleic acid sequence-based amplification (NASBA) and real-time RT-PCR for detection of noroviruses in faecal material. J Virol Methods 135:163–172
123. Liu C, Geva E, Mauk M, Qiu X, Abrams WR, Malamud D, Curtis K, Owen SM, Bau HH (2011) An isothermal amplification reactor with an integrated isolation membrane for point-of-care detection of infectious diseases. Analyst 136:2069–2076
124. Rodriguez NM, Linnes JC, Fan A, Ellenson CK, Pollock NR, Klapperich CM (2015) Paper-based rna extraction, in situ isothermal amplification, and lateral flow detection for low-cost, rapid diagnosis of influenza a (H1N1) from clinical specimens. Anal Chem 87:7872–7879
125. Vincent M, Xu Y, Kong H (2004) Helicase-dependent isothermal DNA amplification. EMBO Rep 5:795–800
126. Piepenburg O, Williams CH, Stemple DL, Armes NA (2006) DNA detection using recombination proteins. PLoS Biol 4:e204
127. Mondal D, Ghosh P, Khan MAA, Hossain F, Böhlken-Fascher S, Matlashewski G, Kroeger A, Olliaro P, Abd El Wahed A (2016) Mobile suitcase laboratory for rapid detection of Leishmania donovani using recombinase polymerase amplification assay. Parasit Vectors 9:281
128. Renner LD, Zan J, Hu LI, Martinez M, Resto PJ, Siegel AC, Torres C, Hall SB, Slezak TR, Nguyen TH, Weibel DB (2017) Detection of ESKAPE bacterial pathogens at the point of care using isothermal DNA-based assays in a portable degas-actuated microfluidic diagnostic assay platform. Appl Environ Microbiol 83:e02449–e02416
129. Teoh B-T, Sam S-S, Tan K-K, Danlami MB, Shu M-H, Johari J, Hooi P-S, Brooks D, Piepenburg O, Nentwich O, Wilder-Smith A, Franco L, Tenorio A, AbuBakar S (2015) Early detection of dengue virus by use of reverse transcription-recombinase polymerase amplification. J Clin Microbiol 53:830–837
130. Daher RK, Stewart G, Boissinot M, Bergeron MG (2016) Recombinase polymerase amplification for diagnostic applications. Clin Chem 62:947–958
131. Cordray MS, Richards-Kortum RR (2012) Emerging nucleic acid-based tests for point-of-care detection of malaria. Am J Trop Med Hyg 87:223–230
132. Green AA, Silver PA, Collins JJ, Yin P (2014) Toehold switches: de-novo-designed regulators of gene expression. Cell 159:925–939
133. Pardee K, Green AA, Ferrante T, Cameron DE, DaleyKeyser A, Yin P, Collins JJ (2014) Paper-based synthetic gene networks. Cell 159:940–954
134. van der Meer JR, Belkin S (2010) Where microbiology meets microengineering: design and applications of reporter bacteria. Nat Rev Microbiol 8:511–522
135. Slomovic S, Pardee K, Collins JJ (2015) Synthetic biology devices for in vitro and in vivo diagnostics. Proc Natl Acad Sci U S A 112:14429–14435

136. Ostrov N, Jimenez M, Billerbeck S, Brisbois J, Matragrano J, Ager A, Cornish VW (2017) A modular yeast biosensor for low-cost point-of-care pathogen detection. Sci Adv 3:e1603221
137. IEA (2016) World Energy Outlook. https://www.iea.org/weo/newsroom/news/2016/november/world-energy-outlook-2016.html
138. IEA (2014) Africa Energy Outlook, a focus on energy prospects in sub-saharan Africa [WWW Document]. URL https://www.iea.org/publications/freepublications/. Accessed 7 June 2017
139. GMSA Intelligence (2017) GMSA Intelligence accesses June 2017 at https://www.gsmaintelligence.com
140. Ventures Africa (2015) 60% OF THE WORLD'S POPULATION WILL OWN A MOBILE PHONE BY 2020 accessed June 2017. http://venturesafrica.com/60-of-the-worlds-population-will-own-a-mobile-phone-by-2020/
141. World Bank (2017) Rapid Urbanization in Africa: Investing in the Development of Africa's Cities [WWW Document]. World Bank. http://www.worldbank.org/en/news/feature/2017/05/02/rapid-urbanization-in-africa-investing-in-the-development-of-africas-cities. Accessed 7 June 2017
142. Lafleur LK, Bishop JD, Heiniger EK, Gallagher RP, Wheeler MD, Kauffman P, Zhang X, Kline EC, Buser JR, Kumar S, Byrnes SA, Vermeulen NMJ, Scarr NK, Belousov Y, Mahoney W, Toley BJ, Ladd PD, Lutz BR, Yager P (2016) A rapid, instrument-free, sample-to-result nucleic acid amplification test. Lab Chip 16:3777–3787
143. Yager P, Domingo GJ, Gerdes J (2008) Point-of-care diagnostics for global health. Annu Rev Biomed Eng 10:107–144
144. Martinez AW, Phillips ST, Whitesides GM (2008) Three-dimensional microfluidic devices fabricated in layered paper and tape. Proc Natl Acad Sci U S A 105:19606–19611
145. Wang C-C, Hennek JW, Ainla A, Kumar AA, Lan W-J, Im J, Smith BS, Zhao M, Whitesides GM (2016) A paper-based "pop-up" electrochemical device for analysis of beta-hydroxybutyrate. Anal Chem 88:6326–6333
146. Zhou G, Mao X, Juncker D (2012) Immunochromatographic assay on thread. Anal Chem 84:7736–7743
147. Choi K, Ng AHC, Fobel R, Wheeler AR (2012) Digital microfluidics. Annu Rev Anal Chem (Palo Alto, Calif) 5:413–440
148. Srinivasan V, Pamula VK, Fair RB (2004) An integrated digital microfluidic lab-on-a-chip for clinical diagnostics on human physiological fluids. Lab Chip 4:310–315
149. Mousa NA, Jebrail MJ, Yang H, Abdelgawad M, Metalnikov P, Chen J, Wheeler AR, Casper RF (2009) Droplet-scale estrogen assays in breast tissue, blood, and serum. Sci Transl Med 1:1ra2
150. Ng AHC, Lee M, Choi K, Fischer AT, Robinson JM, Wheeler AR (2015) Digital microfluidic platform for the detection of rubella infection and immunity: a proof of concept. Clin Chem 61:420–429
151. Fobel R, Kirby AE, Ng AHC, Farnood RR, Wheeler AR (2014) Paper microfluidics goes digital. Adv Mater 26:2838–2843
152. Dixon C, Ng AHC, Fobel R, Miltenburg MB, Wheeler AR (2016) An inkjet printed, roll-coated digital microfluidic device for inexpensive, miniaturized diagnostic assays. Lab Chip 16:4560–4568
153. Fobel R, Fobel C, Wheeler AR (2013) DropBot: an open-source digital microfluidic control system with precise control of electrostatic driving force and instantaneous drop velocity measurement. Appl Phys Lett 102:193513
154. Güder F, Ainla A, Redston J, Mosadegh B, Glavan A, Martin TJ, Whitesides GM (2016) Paper-based electrical respiration sensor. Angew Chem Int Ed Engl 55:5727–5732
155. Conrad CC, Hilchey KG (2011) A review of citizen science and community-based environmental monitoring: issues and opportunities. Environ Monit Assess 176:273–291

156. Hamedi MM, Campbell VE, Rothemund P, Güder F, Christodouleas DC, Bloch J-F, Whitesides GM (2016) Electrically activated paper actuators. Adv Funct Mater 26:2446–2453
157. Safavieh R, Juncker D (2013) Capillarics: pre-programmed, self-powered microfluidic circuits built from capillary elements. Lab Chip 13:4180–4189
158. Jenum S, Dhanasekaran S, Lodha R, Mukherjee A, Saini DK, Singh S, Singh V, Medigeshi G, Haks MC, Ottenhoff THM, Doherty TM, Kabra SK, Ritz C, Grewal HMS (2016) Approaching a diagnostic point-of-care test for pediatric tuberculosis through evaluation of immune biomarkers across the clinical disease spectrum. Sci Rep 6:18520

Chapter 7
The Fascination of Paper

Jan G. Korvink

Abstract Paper is a medium that, despite its long history, doesn't cease to fascinate. When some unrelated technology is eventually transferred to a paper basis, human beings respond with wonder and amazement. Thus we find that people often are prepared to invest much personal effort to help facilitate this step. In some cultures, notably in China and Japan, paper is traditionally treated with great respect, and to this day paper crafts are maintained to a high degree of sophistication. The respect afforded to paper applications is certainly no different for paper-based medical diagnostics. One might consider the cause of this fascination. Paper is certainly ubiquitous, and most likely strongly linked to our childhood experiences, so that most people actually possess fairly considerable paper skills and intuitions by the time they reach into adulthood. But more than that, we observe that paper is so uncanningly versatile, which encourages us to seek new applications beyond the obvious. In this contribution we take a look at paper from this perhaps unusual perspective, complementing the more detailed scientific viewpoints in the remainder of this book.

7.1 Introduction

In 1967, Time magazine ran a headline article entitled "*Fashion: Real Live Paper Dolls*" in which a story was told about the San Francisco department store Joseph Magnin Co. selling disposable and fashionable paper clothing at 1$ a dress. As a 6-year-old child growing up on a small farm in far away South Africa, I heard my parents discussing this story with awe and wonder. Now, I of course knew paper clothing, because popular magazines at the time ran children's sections with printed cutouts for dolls, and my sisters had immense collections of these, all of the most current of fashions. However, what the Time Magazine was reporting was totally

J. G. Korvink (✉)
Karlsruhe Institute of Technology, Karlsruhe, Germany
e-mail: jan.korvink@kit.edu

K. J. Land (ed.), *Paper-based Diagnostics*,
https://doi.org/10.1007/978-3-319-96870-4_7

different, because to suit the needs of real people, the paper clothing was made malleable and tough, so that it would not tear in use. In the 'sixties unplanned nakedness would have been a tremendous scandal and a great barrier to usage. Indeed, for my parents' generation, the radical idea was first and foremost the new fact that clothing could be single use (how wickedly wasteful for the postwar generation). Of course in the 1960s, paper was one of the few materials already considered disposable worldwide, and so in fact contributed to the acceptance of this idea. Paper of course was not on equal footing with cloth, so that it was also tremendously newsworthy that wearing paper would be both safe, and that it would also "feel" as if one was wearing cloth.

In our childhood, we encounter paper often, and have much time to experience its fascination, and ponder its uses. To quote [1]

> *'We admired the spontaneity of real and unreal uncles, who celebrated the emergence from a simple sheet of paper, a hat, a boat, an airplane, using complicated folding processes. Questionably blessed with school books, we later learned to form the airplanes ourselves, a toy that did not lose its desirability even in the highest institutions of learning. Less harmless were the triangular projectile friends, tips equipped with rusty nibs: they remained, elegantly vibrating, stuck in the schoolroom doors.'* (Buchmann, 1969)

The aspect of paper that makes it an ideal feedstock for the production of disposable items appears to be already 60 years older. According to the Smithsonian, paper plates for food were invented in 1904 (the same year as the publication of Einstein's paper on the special theory of relativity!), followed by paper cups and cutlery in the ensuing years. Indeed, disposability has played a major role in almost all recent and unusual applications of paper, but when considered in the context of the entire history of paper, this is a fairly recent development. Thus disposable paper items are available as napkins and tissues for wiping away liquid and dirt, as post-it notes and notepads for temporary memory, as envelopes for temporary privacy and protection, or as bags for the temporary transport of items of value. Edible versions of paper, for example for baking cakes, or for oblate offerings used in religious ceremonies, appear to have been first made in China from the dried pith of *Tetrapanax papyrifer*, and sold as cheap sweets to children in Europe during the 'sixties.

Paper is so completely woven into our linguistic landscape that the terminology associated with paper is not only rich and varied, but has also been absorbed into numerous idiomatic expressions. The word "*paper*" can variously refer to newspaper, scholarly article, or the material itself. Particularly colourful vocabulary around the topic of paper includes the terms foil, folio, lamina, leaf, octavo, sheet, stripe, roll, stack, ream, and so on. Why has this happened? The role of paper in the spreading of ideas, and hence the importance of its technology in forming affluent cultured societies, is indisputable, and has accompanied us for many centuries. And, because the leading position of paper in this business is now strongly under attack, due to the spread of digitalisation, it is useful to remind ourselves of its history, before we proceed to consider some of the other excellent features of this ultimately versatile medium.

7.2 Early History

Paper has played an extraordinarily important role in the development of civilized culture. Tablets of all kinds (stone, wood, bamboo) dominated as recording medium for officialdom, but were unhandy. In China a breakthrough came at around 105 C.E., when the higher civil servant Ts'ai Lun managed to form thin dried sheets from a mash of waste fibres, yet at first the idea was treated with suspicion. Because the Chinese carefully heeded this secret, it took hundreds of years before the idea spread to Europe, and it was only in the eighth century it had reached Japan [1], a neighbouring country!

In their geographically imposed role as liaison between East and West, the Arabs brought paper making to Europe, so that in the twelfth century the first paper mill was established in Spain. The idea then spread throughout the continent, reaching Basel in 1433, and Zurich in 1470 [1].

Prior to the emergence of the digital era and its alternatives, paper was really the only scalable medium on which to record ideas for prosperity, first in the form of written or printed books, and as scientific endeavour became more organised also in the form of scientific articles, so aptly termed *papers*.

The entire terminology surrounding paper got absorbed into everyday life to the extent that the original meaning evolved considerably. This we talk of *being on the same page* when acknowledgement for a common understanding is seeked, presumably because we can so clearly envisage reading a precise instruction from a page in the book, and if our counterpart is on the same page we have reached the same level of understanding. Or *tabula rasa* to refer to the unwritten page, whenever we have reached a limit in a discussion or negotiation, and want to start all over again.

The making of paper has evolved independently in numerous and considerably separated cultures. The most prominent of these are China, Japan, India, and Egypt, but because of lack of detailed information it is hard to determine exactly the priority. Early papermaking was strongly tied to agricultural finesse, since the feedstock of paper is predominantly plant biomass, such as rice, bamboo, papyrus, or trees. A radical transformation occurred when papermaking became industrial, for access to handmade paper was essentially a privilege of the wealthier classes, so that any associated creativity (painting, drawing, writing) was also stunted. As soon as this monopoly was broken, the demand for paper could rapidly grow, which required not only better and more automated methods of papermaking, but also access to the resources, such as cotton and wood for pulp, and water for processing and as an energy supply.

In early European industrial culture, cities on rivers and close to a supply of timber (Basel, at the foot of the Black Forest, and on the banks of the Rhine, is a prime example, as is Freiburg, Heidelberg, and Mainz) could play their local advantage to produce paper from pulp, and establish a book and journal publishing industry based on a vast array of technical innovations including the penultimate establishment of the printing press by Johannes Gutenberg in 1450. Humanity's insatiable hunger for information was fed by these inventions, and created much wealth. Increased sophistication in these communities led to a paired growth in the academic tradition, and it is no surprise that where printing presses were rolling out books, intellectuals were needed to aid in quality control, and interface with a growing tradition in writing.

Already quite early on in its history, paper was made available in different qualities. Because of its mode of production using biomass, alkaline chemicals, and fresh river water, paper was delivered from the production site in a fairly sterile format, so that it was identified early on for the safe wrapping of food, such as raw meat at the butcher, and fresh bread at the bakery. More noble silky papers were used to wrap expensive clothing, and decorative paper could be used for gifts, to replace cloth which costs more effort to produce.

7.3 Value, Authenticity, and Commerce

It did not take long for paper to emerge with value. Exceptional books have always been of value, but paper also turned out to be the ideal medium to regulate commerce at the more abstract level. Thus, when personal wealth could no longer be practically mapped to the weight of noble metals, its abstraction in the form of printed company shares, letters of debt, property deeds, and bank-guaranteed notes became commonplace. This necessitated the invention of the watermark and other means to make forgery difficult, and methods to make paper more durable, especially from fouling in dampness, so as to avoid the loss of value. To this day, we recognise the special place of paper in representing authenticity and value, in the form of certificates, identification documents such as passports, and of course money.

Perhaps nowhere else than in the manufacture of money has paper technology become so advanced as in the endeavour to achieve authenticity. Today bank notes, still considered important despite the onslaught of digital bitcoins, contain a vast array of measures to protect against forgery. By varying the density of the paper pulp prior to drying, a watermark is embedded in the paper. Often paper money contains a thread as well. The threads may themselves be complex materials that are hard to forge, and may include magnetic and fluorescent markers. Threads are sometimes woven through the pulp, nearing and receding from one face of the paper in a regular and complex pattern. Banknotes already early on played with incorporating metal features, which required sophisticated bonding technologies between the pulp material and the metallic thin film. The metal not only added shine, but could also be used to carry a nanoscale pattern encoding a hologram or kinegram, again features that are hard to copy. Some bank notes also contain transparent windows, usually polymeric materials that have excellent mechanical and optical properties, and furthermore facilitate bonding with the pulp constituents. Till today though, most counterfeit measures are achieved by the printing process, in a sophisticated interplay between pattern, ink, and resolution. Taking the Japanese Yen banknotes as an example, ink topography via intaglio printing is used to form Braille patterns. The watermarks are graded and hence hard to copy, and appear as images as well as machine-readable patterns. Holograms modify the printed images with decorative patterns. Tilting the viewing direction leads to a variety of appearing patterns, both of simplifying and complexifying nature. Also, colour changes are so affected, for example the appearance of pink bands, additionally achieved by colour-shifting inks in the numbers. For machine checking, microprinting and luminescence are used in complex patterns that are hard to copy. Swiss banknotes add a few more measures, such as microperforation in the numbers, microregistration of back and front pattern discernable from both sides, and metallic micropatterns in the numbers.

All of these measures serve to indicate the exceptional versatility and durability of paper to maintain value, and to accept a large variety of modifications and integration with other materials and manufacturing processes.

7.4 Political and Cultural Aspects

Paper has played a key role in the development and establishment of political ideas. The writing down of religious and political laws in books formed an important foundation for the organisation of society, followed by their translation into the vernacular in response to an increasing awareness of the need for democratic access (or pressure by the population to achieve access). Laws were first carved into stone tablets (the Ten Commandments), and then appeared in paper form. It is hard to imagine laws that are only available in electronic form.

Up until broad access to the Internet has been generally available, paper has been the most important carrier and record of political ideas, and as a result the writing and printing of ideas on paper have been subject to repression, secrecy, and excitement. In the postwar period since 1945, lower and middle class families were proud of their *up-to-date* encyclopaedias, thereby offering their growing-up children a peek of universal knowledge in the hop of improving the lot of their lineage.

Dissent with the ruling classes has often also been expressed in pamphlet or poster form, and its printing has equally often been banned, see for example the 1968 youth movement in Paris [2]. During recent wars, the printed pamphlet, dropped from the sky by aeroplane, has regularly been used by an opponent to broadly inform the public. Whenever a new political philosophy needed to be promulgated, it was printed in the form of a manifesto and spread among the population. During preceding cultural and political revolutions, the struggle was often expressed in the form of a transition of the control over information flow, which has often meant control over the printing press for the production of newspapers and pamphlets. Indeed, dictators generally burn liberal books as potentially threatening, and they oftentimes create books as propaganda so as to spread their own philosophy (*Little Red Book* by Mao Zedong, *The Road to Power* by Joseph Stalin, *Mein Kampf* by Adolf Hitler, *Zabiba and the King* by Saddam Hussein, and so on). Paper also underlines a ruler's power. Thus present-day Donald Trump issues his executive orders and proclamations (such as order number 13772 *Core Principles for Regulating the United States Financial System*) by ceremoniously signing a paper version in front of the camera using an oversized signature.

On the cultural side, paper had a similar role to for example render literature, plays, and poems accessible to the general classes, and not just to the educated elite. A significant breakthrough was achieved in the mid-twentieth century, with the paperback business model, in which a particular book was produced in such vast quantities (millions) so as to reduce production costs to a level that makes books accessible to "everyone". Thus a small budget could establish a significant library of format, and popular novels further invigorated the book production industry. Significantly, the monopoly of publishing houses and their editors in deciding content was eventually broken, so that anyone with an idea (regardless of inherent value) could produce and proliferate a book.

The feedstock for this industry is paper, ink, and glue, and judging from the quality of old paperbacks the weak point was definitely the glue. The paper for the book covers, and for the inner pages, has lasted exceptionally long considering the low cost of acquisition.

7.5 Games, Toys, and Leaving Flatland

The continuous industrialisation that has followed out from the first industrial revolution has been interrupted by numerous wars. Yet during these turbulent times, the concept of an *ideal* childhood has nevertheless co-evolved to its current form, with time for playing and the accompanying toys seen as important constituents of the development of a child. Thus, the invention of toys, also during economically strained periods, has benefited strongly from the existence of paper technology. Apart from playing cards and board games, paper pulp has been formed into malleable masses, and turned into jewellery, toy landscapes, and model building bricks to name the most prominent. To this day, on a windy day in Shanghai, peasant vendors sell colourful kites made of paper, thin slats of wood, and string swirling in the clear sky to lure their new owners.

Nevertheless, no single culture has contributed more to the joy and fascination of paper as the land of the rising sun.

The Japanese paper folding technique of *origami* appears to have been around since at least the sixth century. From its very serious roots, it has evolved into a very popular Japanese pastime, even today, and is a great source of inspiration. Its related form, *kirigami*, which involves also paper-cutting steps, provides additional degrees of freedom, especially with respect to the amount of paper needed to arrive at a particular shape, and of the possible kinematics that can be achieved.

Paper folding has also made it into the industrial landscape [13–16]. Using kirigami techniques, paper can be formed into perfect transport packaging that is both fun to unwrap and protective of the contents [3]. It is remarkable that the scientific literature regularly reports on advances based on these age-old techniques (Web of science lists 1279 papers with *origami* and 80 with *kirigami* in the title), the most cited being from science [4].

7.6 Paper in the Arts

Nowhere else is paper used so creatively as in the arts. Through its varying surface texture, paper is conveniently available in a vast array of qualities, with many "allotropes" directly targeting specific artistic techniques. Rough papers are especially popular for drawing with charcoal, crayon, and pencil, mainly because slight variations in drawing pressure can have dramatic effects on the amount of colour taken up by the paper, and on the resulting pattern formed, but of course also because charcoal adheres well to paper. Presumably, sympathetic drawing material chemists have been long conscious of the needs of artists, spurning on endless creativity.

The inner structure of paper plays an enormous role also in watercolour paper, since it governs the principles of water uptake and drying patterns, a phenomenon which has been tremendously explored over the centuries. Anyone working with paper-based diagnostics is well advised to study the techniques of the masters, such as the impressionist Paul Cézanne, whose pigmented inks are manipulated to form variously sharp and diffuse edges, and gentle layering to achieve sheer unlimited colour and shading effects in and on the paper [5].

Over the years, art academies have lost their influence, leading to the dropping of conventions in art, allowing paper to be explored also for its potential to achieve a vast variety of textures and messages by other means than painting and drawing. Thus folding, crumpling, and tearing, combined with piercing, embossing, moulding, and burning, have easily entered the vocabulary of artists. Indeed, hardly a technique is not used, combined, layered, and stacked to achieve the creator's final goal.

7.7 Paper as a Working Material

Everyone might remember their first cut due to a carelessly handled sheet of paper, and the surprise at discovering that such a "soft" material could cause so much hurt and discomfort. Or at the opposite extreme, we enjoy the softness on the skin of a good-quality tissue paper, or its comforting drying effect when we need it. Nowadays, we package fragile items in paper, using its various qualities to absorb impact, provide a crunch zone, protect from moisture such as rain, and standardise the shape of packaging for ease of handling. Indeed, how would the Internet dealer Amazon manage without paper?

When we list popular adjectives for materials, we discover that many of the terms can easily apply to various generally available paper types. Thus paper is resilient, strong, weak, soft, hard, sharp, smooth, shiny, rough, opaque, transparent, permeable, absorbent, abrasive, insulating, conductive, etc.

Similarly, we can investigate the various ways in which paper can be transformed. Paper can be cut, lasered, sawn, torn, poked and punched, folded, textured, rolled, laminated, wrinkled, stiffened, graded, shaped, moulded, glued, welded, bonded, ironed, polished, bound, spun, sewn, woven, printed, painted, sprayed, etc. [11–18]. Paper money is even regularly washed and sterilized.

As a result, and despite the fact that many conventional tools and machines can be used to work paper, a large industry has evolved that makes paper machining more efficient and precise [15, 16]. We do not live in the heyday of this industry, which is already diminishing in size and has become highly specialised. Nevertheless, a clear survivor is the packaging industry. At the small size end of the business, machines exist that print and fold individualised letters, insert them into subsequently sealed envelopes, and correctly sort them at post offices around the world. Almost everything we consume in bulk has been individually packaged, from cosmetics, through food, to luxury items. And wherever possible, paper in one of its many forms plays a dominant role, customised for function and aesthetics. One can only speculate that paper has retained its exceptional leading position in cost versus utility.

7.8 Paper in Mathematics and the Natural Sciences

We commonly use paper as an analogy for conceptualising about two dimensions, treating the sheet of paper as if it has no thickness. This is because in our common experience we live in an essentially 3D world; a 2D world cannot sustain life at all. In 1884 the English school teacher Edwin Abbott wrote a fantasy of a 2D world called Flatland [6], in which he explored the conditions for existence without a third dimension. Subsequently other authors also explored the idea, including Dr. Steven Hawkin who notes in [7]:

> *... the impossibility of life in two-dimensional space, as any inhabitants would necessarily be unable to digest their own food.* (Hawkin, 2001)

In English we speak of *paper thin* to indicate this vanishing thickness. And so, whenever a two-dimensional thought experiment needs to be performed, a sheet of paper is the obvious choice.

The origami folding of a two-dimensional sheet discussed earlier is a mathematical problem domain with particularly rich structure. For example, a finite polygonal sheet of paper, subjected to folding, contracts its outer perimeter into a new polygonal shape. What is the minimal number of folds required to achieve a particular polygonal perimeter? The crumpling of a finite sheet of paper in situ leads to a theorem which claims that at least one point in the crumpled sheet occupies the same two-dimensional coordinates as it did in the sheet. Thus the science of origami has grown strongly in recent years, and even features a large popular following despite the depth of theory. Robert Lang deserves special mentioning, because he created the Mathematica® program Tessalica® with which the folding pattern of a sheet of paper, aimed to achieve a particular 3D model shape, can be discovered automatically [8]. Another leading scientist is Eric Demaine, who has proven that any polygonal shape can be obtained from a multiply folded sheet of paper followed by a single straight-line cut [9].

The *litmus test* is a phrase that has entered everyday speech to infer a simple test that reveals the absolute truth about something. The word comes from Old Norse (litmosi), referring to a moss used for dying. Indeed, in science, litmus paper contains a dyestuff (extracted from lichens) that changes colour upon exposure to a solvent with pH above a threshold. It has become a paradigm for the paper microfluidics community. Paper is a carrier of both dye and controlled fluidic environment, wicking up the liquid and ensuring good contact with the analyte. The colour change is vivid, and easy to judge. In a subsequent sophistication, the *lateral flow test* embeds an architecture that arranges different analytes in stripes across the flow direction of the strip of paper. The liquid (blood, sputum, urine, etc.) is carried along the strip to pass the analytes in a controlled sequence and at a controlled rate, allowing time for reactions to take place. The numerous excellent chemical and physical properties of paper have greatly facilitated these laboratory applications, which continue to evolve. In the form of a pregnancy test, this simple strip of paper decides over fates.

In science, the word *paper* is synonymous with *knowledge*. The scientific community attempts, through peer review, to improve the truthfulness and relevance of ideas expressed in approved articles called *papers*. These records, interlinked with each other through cross-references, are our most important repositories of knowledge. If I want to learn about Emmy Noether's ideas, I can read her papers [10]. Up until the early 1990s, journals would offer their paper authors paper reprints, neatly bound copies of the paper (usually 10–20 pages long) for distribution to collaborators and friends. Thus the paper, to the outsider a grain of knowledge within a web of ideas, is also a unit of scientific productivity (how many papers this year?) and quality (in which journal did you publish?).

7.9 The Continuous Reinvention of Paper

The paper industry has always played an important role in industrialisation, generating many jobs and solid income for investors [17, 18]. Especially in the following countries, China, the USA, Japan, Germany, Korea, Brazil, Finland, Canada, Sweden, and Italy, the coexistence of larger rivers and forests has led to a thriving paper production industry, but also to the necessity to continuously invent new roles for paper. This is because every major technological advance, but especially that in digitalization, has encroached upon paper's prominence in the marketplace. Furthermore, affluent countries are continuously changing legislation surrounding recycling procedures, leading to the need for adaptation in the industry, to accommodate these modalities and remain competitive.

Thus paper's prominent survival has been built upon an agile industry that has managed to stem these "onslaughts", but whereas, from today's perspective, paper will not disappear, it is losing ground.

7.10 The Replacement of Paper

Despite our fascination for paper, and our sentimentality for its role in the past, paper's position has never been so challenged as during the recent two decades. Almost all of paper's central roles are being overtaken by other technologies. One of the most prominent attaches on paper's supremacy is in the gradual decline of the book as the most important conveyor of information. The trend is driven by the accessibility of the Internet and the ease with which large informational resources can be searched at low cost, but also by the convenience of digital reading devices (yes, reading is still important).

In 1976–1984, John Warnock established the revolutionary idea that a printed page could be conveniently described by a universal programming language (PostScript®, commercialised by Adobe® Systems) that did not depend on the resolution of the printer, but only on the intention of the page creator. Once printing machines adopted this idea, it became possible to print high-quality pages in the office and at home, all at once democratising the printing industry. In hindsight, this was probably one of the most important advances since the invention of the printing press in 1450, because it was a technical innovation and also introduced a change in roles. Although it has not made printers completely redundant, it removed the inherent control over the printed format.

Based on the PostScript® language, Adobe® then invented the Portable Document Format® (PDF) that embedded PostScript® commands in a format that made digital books possible, in the sense that digital books can be browsed, searched, and, importantly, read as if they were printed. Another advantage, owning a vast library of PDF books, adds no significant weight to a portable reader device, but adds many more useful functions, including rapid searching. PDF has made it possible to share complex documents in brilliant quality, and has been continuously extended to add features useful in digital commerce and government, such as needed to ensure authenticity, or the embedding of other media forms in the document, such as 3D graphics, or sound. This advance, so necessary for our modern lives, has contributed significantly to the demise of paper by replacing and improving many of its functions. As is so typical, many of the other useful functions are being sacrificed because values change, and convenience dominates human attitudes.

Paper is known in technical circles as a non-woven. Its structure is based on 5–10 μm diameter cellulose fibres arranged in a random but homogenised structure (in the plane of the sheet) and layered in the thickness direction: a typical sheet of paper has at least two more or less smooth outer layers, and a fairly homogenous interior. Many applications require more controlled inner structure, which will of course lead to improved paper-like functions. Thus the natural celluloses are gradually being replaced by other polymer fibres, and inner structure at the micro- and macroscale is being achieved by advances in material processing. Also the macroscale properties are being modified to achieve exceptional toughness and strength (materials such as Tyvek® and Teslin®), and resistance to a variety of chemicals and radiation, and retain good qualities for processing.

7.11 Postscript

In the future, traditional paper might 1 day shrink in importance, until only maintained by a small community rooted in the arts and crafts, where randomness and surprise are wanted features that drive creativity and exploration. Since these remain important driving forces also in science and technology, we are well advised to continue also our explorations of paper for an ever-widening range of applications.

Acknowledgements I gratefully acknowledge the following agencies for their continued support of my work: the Deutsche Forschungsgesellschaft (DFG), especially for the grants Rolled up microsystems (KO 1883/23-1), MetaCoils (KO 1883/20-1), and BrainLinks-BrainTools (EXC 1086), and the European Research Council (ERC), especially for the grants NMCEL, LiVOX, and LockChip. I also got benefit from the excellent research infrastructure at the Karlsruhe Institute of Technology, which I hereby acknowledge.

References

1. Buchmann M (1969) Magie des Papiers. Exhibition in the Kunstgewerbemuseum Zürich. 29 November 1969 to 25 January 1970. Zürich, Switzerland
2. Damamme D, Gobille B, Matonti F Pudal B (eds) (2008) Mai-juin 68 (in French). E´ ditions de l'Atelier
3. Denison E, Cawtray R (1999) Packaging prototypes. RotoVision SA, Mies
4. Han DR et al (2011) DNA origami with complex curvatures in three-dimensional space. Science 332(6027):342–346
5. Doran M (ed) (1991) Gespräche mit Ce'zanne [Conversations with Ce'zanne]. Verlag, Diogenes
6. Abbott EA (1884) Flatland: a romance of many dimensions. Seeley & Co, Phoenixville, PA Reprint ISBN 1522968466
7. Richard GJ (2001) Time travel in Einstein's universe: the physical possibilities of travel through time. Boston, MA, Houghton Mifflin Company, p 61 ISBN 978-0395955635
8. Robert J (2017) Lang Origami. http://www.langorigami.com/article/tessellatica
9. Demaine E et al (2001) Polygons cuttable by a circular saw, computational geometry: theory and applications. Comput Geomet Theory Appl 20(12):69–84 Special issue of selected papers from CCCG 2000

10. Noether E (1908) Über die Bildung des Formensystems der ternären biquadratischen Form [On complete systems of invariants for ternary biquadratic forms], Journal für die reine und angewandte Mathematik (in German), DE: Uni Göttingen, 134:2390
11. Avella N (2003) Paper engineering. 3D design techniques for a 2D material. Rotovision SA, Mies ISBN 2888930498
12. Croy P (1964) Grafik. Form + Technik. Musterschmidt, Göttingen. ASIN B0053TEO4G
13. Jackson P (2011) Folding techniques for designers: from sheet to form. Laurence King Publishing, London ISBN 1856697215
14. Jackson P (2012) Structural packaging: design your own boxes and 3D forms. Laurence King Publishing, London ISBN 1856697533
15. Jackson P (2013) Cut & fold techniques for promotional materials. Laurence King Publishing, London ISBN 178067094X
16. Jackson P (2014) Cut and fold techniques for pop-up designs. Laurence King Publishing, London ISBN 1780673272
17. Williams N (1993) Paperwork, Phaidon Press Ltd., London. ISBN 0714834610
18. Zeier F (2013) Papier, Haupt, 6th ed. ISBN 3258012059

Index

A

Actuators and control modules
  cellulose and hybrid materials, 134
  heating components, 135–136
  high-performance paper actuators, 134
  membrane switches, 134
Affordable, Sensitive, Specific, User-friendly, Rapid and Robust, Equipment-free and Deliverable (ASSURED) criteria, 2, 8, 9, 15, 19, 191
  actuators and control modules, 124, 134–136
  automated communication modules, 125
  automated read-out, 125
  components, 159–161
  computational capabilities, 125
  connectivity modules, 124, 148–151
  data processors and storage modules, 124, 140–142
  displays and communication units, 125
  energy storage components, 124, 151–158
  integration, 124, 159–161
  low-resourced settings, 125
  portable potentiostat systems, 124
  read-out and displays, 124
  sensors, 124
  WHO ASSURED criteria, 2–5, 181
Alere Pima analyser, 15
Antibodies, 181–183, 185
Antibody-based detection, 64–65
ASSURED criteria, *see* Affordable, Sensitive, Specific, User-friendly, Rapid and Robust, Equipment-free and Deliverable criteria

B

Biofuel cells, 154
Biosensing applications, 127
Black Lumocolor®, 111
Business and economic models
  Alere Pima analyser, 15
  clinical environment, 15
  community health workers (CHWs), 15
  cost saving aspects, 16
  Diagnostics for All (DfA), 16
  GeneXpert system, 15
  healthcare funding models, 15
  insurance and transport, 15
  population growth and unemployment, 14
  TB screening and HIV tests, 17
  WHO CHOICE program, 17

C

Capacitors, 131–132
Cellulosic material, 40
Cepheid GeneXpert technology, 177
Chemical pulping, 25
Chemiluminescence, 142
Colorimetric detection
  digital camera, 67
  image analysis, 68
  mobile phone camera, 67–68
  qualitative outputs, 72–75
  scanner, 67
Community health workers (CHWs), 15
Connectivity modules
  printed electronics, 149–151
  wireless communication technologies, 148–149

K. J. Land (ed.), *Paper-based Diagnostics*,
https://doi.org/10.1007/978-3-319-96870-4

D

Data processors and storage modules
  integrated circuits, 140–141
  memory components, 141–142
Diagnostics for All (DfA), 16
Digital camera, 67
Digital microfluidics (DMF), 189
Digital readers
  handheld readers, 75–78
  smartphone-based readers, 78–79
Digital-to-analogue converters (DACs), 140
Disposable paper items, 204
Dry and wet strength additives, 36
  H-bonds, 38
  melamin formaldehyde (MF), 38
  natural polymers, 37
  polyacryl amide-based polymers, 38
  starch, 37, 38

E

Electrochemical batteries, 151–152
Electrochemical detection
  electrode fabrication methods, 69–70
  quantitative results, 75
Electrochemical transistors (ECTs), 133
Electrochemiluminescent displays, 146
Electrochromic displays, 144–145
Electrode fabrication methods, 69–70
Electroluminescent displays, 144–145
Electrophoretic displays, 147
Energy storage components
  biofuel cells, 154
  electrochemical batteries, 151–152
  lithium-ion batteries, 153
  nanogenerators, 157–158
  solar cells, 157
  supercapacitors, 155–156
Enzyme-based detection, 62–64
Enzyme-linked immunosorbent assays (ELISAs), 51, 171

F

Fabrication methods, 100–101
Fast Lithographic Activation of Sheets (FLASH), 102
Fiber fractions, 26
Fiber type, role of, 40–42
Filtration isolation of nucleic acid (FINA), 179
Flexography printing, 108–110
Flow manipulation tools
  automated methods, 52
  dielectrics, 52
  magnetic switches, 52
  microfluidic paper-based analytical devices (μPADs), 47
  paper modification, 52
  spot tests, 47
  switches and valves, 50–52
  wicking speed, 48–50
Fluidic manipulation
  sample distribution and multiplexing, 90–91
  unit operations
    mixing, 94–95
    timers and fluidic interrupters, 94
    valve system, 93–94
Forensic analysis, 62
Fractionation, 27–29, 42, 44

G

Gas sensor, 138
GeneXpert system, 12, 15
Glucose testing, 10

H

Handheld readers
  ePADs, 75–76
  external reader, 76–77
  potentiometric measurements, 76, 78
  smartphones, 76
H-bonds, 37, 38
Heat energy, 33
Hemicelluloses, 24, 40
Hexadecyltrimethylammonium bromide (CTMAB)-modified paper, 51
Hollow-channel PADs (HC-PAD), 50
Home pregnancy test, 171
Humidity, 137
Hybrid paper circuits, 129–130
Hybrid printed systems, 140

I

Image analysis, 68
Inclined wire headbox, principles of, 32
Inductors, 131, 132
Inkjet etching, 107–108
Inkjet printing
  hydrophobic UV-curable acrylate, 106
  inkjet etching, 107–108
  mass production, 107
Integrated circuits, 140–141

K
Kirigami, 208
Klemm-Test ISO 8787, 27

L
Lab-on-a-foil processes, 127
Laser treatment, 111
Leapfrog technology, 9
Lewin Group, 1
Light and image sensors, 139–140
Light emitting electrochemical cells, 145
Lignin, 24, 25
Lithium-ion batteries, 153, 155
Litmus test, 211
Luminescence detection approaches, 71

M
Mass spectrometry (MS), 71
Mechanical pulping, 25
Melamin formaldehyde (MF), 38
Microbial water analysis, 13
Microdialysis, 176
Mobile phone camera, 67–68
Multiplexable autonomous disposable nucleic acid amplification test (MAD NAAT), 187

N
Nanogenerators, 157–158
Nanoparticles, 61–62
Natural fibers, 24
Natural polymers, 37
Nucleic acid
  amplification, 65, 183–185
  hybridization, 65–66
  sampling, 174

O
Octadecyltrichlorosilane (OTS), 113
Organic light emitting diodes (OLEDs), 146–147
Origami, 208

P
Packaging materials, 24
Paper
  arts, 209
  definition of, 204
  early history of, 205–206
  games and toys, 208–209
  in mathematics and natural sciences, 210–211
  political and cultural aspects, 207–208
  reinvention of, 211–212
  replacement of, 212–213
  role of, 204
  values, 206
  watermarks, 207
  working material, 209–210
Paper-based diagnostics
  actuators and control modules
    cellulose and hybrid materials, 134
    heating components, 135–136
    high-performance paper actuators, 134
    membrane switches, 134
  connectivity modules
    printed electronics, 149–151
    wireless communication technologies, 148–149
  data processors and storage modules
    integrated circuits, 140–141
    memory components, 141–142
  detection
    colorimetry, 66–68
    electrochemical detection, 68–70, 75
    luminescence detection approaches, 71
    mass spectrometry (MS), 71
    surface-enhanced Raman spectroscopy (SERS), 71
  energy storage components
    biofuel cells, 154
    electrochemical batteries, 151–152
    lithium-ion batteries, 153
    nanogenerators, 157–158
    solar cells, 157
    supercapacitors, 155–156
  integration and components, 159–161
  reaction mechanisms
    antibody-based detection, 64–65
    environmental analysis, metal complexation for, 61
    enzyme-based detection, 62–64
    forensic analysis, 62
    nanoparticles, 61–62
    nucleic acid amplification, 65
    nucleic acid hybridization, 65–66
  read-out and displays
    advantages and limitations, 143
    chemiluminescence, 142
    electrochemiluminescent displays, 146
    electrochromic displays, 144–145
    electroluminescent displays, 144–145
    electrophoretic displays, 147

Paper-based diagnostics (*cont.*)
light emitting electrochemical cells, 145
organic light emitting diodes (OLEDs), 146–147
thermochromic displays, 144
visual read-out approaches, 142
unmet needs, for developing world
business models and economics, 14–17
connectivity, 6
development pipeline, 11–12
device components, 6
disposable and biodegradable properties, 7
environmental concerns, 5
equipment-based diagnostics, 18
fluidic and functional requirements, 9
functional components, 18
functional printable materials, 6
GeneXpert system, 12
glucose testing, 10
healthcare and environmental consciousness, 13
HIV screening, 2
leapfrog technology, 9
long-term deployment plan, 7
machine-operated microfluidic cartridge, 8
microbial water analysis, 13
pH testing, 10
physical and chemical properties, 7
point-of-care diagnostics, 7, 8
polymerase chain reaction (PCR), 7
positive temperature coefficient (PTC), 7
quantitative outputs, 19
resource-limited settings, 9, 13
roll-to-roll printing processes, 9
SWOT analysis, 7, 8
timeline of, 11
urban to rural/clinic settings, 18
WHO ASSURED criteria, 2–5
Paper-based enzyme-linked immunosorbent assay (p-ELISA) microplate, 111
Paper-based microfluidic analytical devices (μPADs)
corona discharge treatment, 113
design criteria
detection zone, 95
fluidic manipulation, 90–95
general consideration, 96–98
hydrophobic barriers, 89
physical limits, 89
sample introduction, 90
3D-μPAD, 96
economics
energy consumption, 113
initial investment, 113
reagent consumption and toxicity, 113
substrates, of hydrophobic barriers, 114–116
fabrication methods, 100–101
flexography printing, 108–110
inkjet printing
hydrophobic UV-curable acrylate, 106
inkjet etching, 107–108
mass production, 107
laser treatment, 111
octadecyltrichlorosilane (OTS), 113
photolithography, 99, 102–103
plasma treatment, 104–105
plotting, 103–104
scalability and cost, 116–118
screen printing, 110–111
stamping, 111–112
wax dipping, 106
wax printing, 105–106
Paper chemical additives, 45
Paper chemistry
dry and wet strength additives, 36
H-bonds, 38
melamin formaldehyde (MF), 38
natural polymers, 37
polyacryl amide-based polymers, 38
starch, 37, 38
functional chemicals, 35
paper-based microfluidic devices, impact of
cellulosic material, 40
fiber type, role of, 40–42
paper chemical additives, 45
porosity, and fiber pretreatment, 42–44
raw material consumption, 35
retention aids, 36–37
Paper clothing, 204
Paper manufacture
applications, 24
conventional papermaking, 24
dynamic handsheet-forming process, 30–31
graphical paper-grade production, 24
industrial papermaking
center roll concept, 33
components, 32
cylinder mold headboxes, 33
heat energy, 33
inclined wire headbox, principles of, 32
mechanical forces and drying, 31
paper web, 34, 35
thermal processes, 33

lignin and hemicelluloses, 24
natural fibers, 24
packaging materials, 24
papermaking, 29
pH determination, 24
pulping and processing
chemical pulping, 25
fiber dimensions, 27
fiber fractions, 26
fractionation, 28
Klemm-Test ISO 8787, 27
mechanical pulping, 25
pine kraft pulp fibers, 27
pressure screens, 27
refining or beating process, 25, 26
screen baskets, 28
slotted screen basket, 29
Rapid-Köthen handsheet method, 29–30
Paper web, 34, 35
Paraffinized paper (p-paper), 111
pH determination, 24
Photolithography, 99, 102–103
pH testing, 10
Pine kraft pulp fibers, 27
Plasma treatment, 104–105
Plotting, 103–104
POCKET immunoassay, 171
Point-of-care (POC) diagnostics, 7, 8
design and implementation, 191–192
equipment-free diagnostics
paper-based devices, 188–189
paper digital microfluidics, 189–190
materials engineering and nanotechnologies, 193
sample preparation
blood collection, 173–174
electrical fields, 180
enzymatic/chemical, 178–179
heat, extraction, 177–178
mechanical lysis, 179–180
respiratory samples, 174
separation process, 175–176
sputum samples, 177
swab-based methods, 175
technologies, 192
urine and secreted fluids, 176
sensitivity and specificity
antibodies, 181–183
nucleic acids, 183–185
Point-of-care testing (POCT) tools, 90
Polyacryl amide-based polymers, 38
Polyamidoamine epichlorohydrin (PAAE) polymeric resins, 38
Polymerase chain reaction (PCR), 7
Positive temperature coefficient (PTC), 7
Pressure screens, 27
Pressure sensor, 138–139
Printed electronic components
capacitors, 131–132
inductors, 132
resistors, 131
structural design, 131
transistors, 132–134
Printing materials
carbon, 128
inorganic and organic materials, 129
semi-conductors and dielectrics, 129
Printing techniques
biosensing applications, 127
electrical and sensing capabilities, 127
electronic structures and functional components, 127
flexible and wearable technologies, 126
hot embossing, 127
lab-on-a-foil processes, 127
low-cost printing solutions, 127
manual sheet screen printing process, 128
roll-to-roll processes, 126
semi-automated sheet, 128
wax screen printing, 127

R

Rapid-Köthen handsheet method, 29–30
Read-out and displays
advantages and limitations, 143
chemiluminescence, 142
electrochemiluminescent displays, 146
electrochromic displays, 144–145
electroluminescent displays, 144–145
electrophoretic displays, 147
light emitting electrochemical cells, 145
organic light emitting diodes (OLEDs), 146–147
thermochromic displays, 144
visual read-out approaches, 142
Resistive sensors, 137
Resistors, 131
Respiratory samples, 174
Roll-to-roll printing processes, 9, 126

S

Sample loading, PAD
sequential delivery, 55–56
single delivery, 53–54

Sample pretreatment
  interference removal, 57–58
  preconcentration, 58–60
Scanner, 67
Screen baskets, 28
Screening test, 1
Screen printing, 110–111
Sensor
  gas sensor, 138
  humidity, 137
  light and image sensors, 139–140
  physical and chemical printed sensors, 136
  pressure sensor, 138–139
  resistive sensors, 137
  temperature sensors, 137
  touch and proximity sensors, 138
Signal amplification
  antibodies
    enzymes, 182
    nanoparticles, 182–183
  nucleic acids
    isothermal amplification, 183–185
    synthetic biology, 185–186
Slotted screen basket, 29
Smartphone-based readers, 78–79
Solar cells, 157
Solid-phase microextraction (SPME), 176
Sputum samples, 177
Stamping, 111–112
Starch, 37, 38
Supercapacitors, 155–156
Surface-enhanced Raman spectroscopy (SERS), 71
Swab-based methods, 175
SWOT analysis, 7, 8
System for nucleic acid purification (SNAP), 179

T
Temperature sensors, 137
Thermal processes, 33
Thermochromic displays, 144
Touch and proximity sensors, 138
Transistor arrays, 140
Transistors, 132–134

U
Urine biomarker, 176

V
Visual read-out approaches, 142
Volatile organic compound (VOC), 108

W
Warnock, John, 212
Wax dipping, 106
Wax printing, 105–106
Wax screen printing, 127
WHO ASSURED criteria, 2–5, 181
WHO CHOICE program, 17
Wireless communication technologies, 148–149
Write once read many (WORM) permanent memory, 142

MIX
Papier aus verantwortungsvollen Quellen
Paper from responsible sources
FSC® C105338

If you have any concerns about our products,
you can contact us on
ProductSafety@springernature.com

In case Publisher is established outside the EU,
the EU authorized representative is:
Springer Nature Customer Service Center GmbH
Europaplatz 3, 69115 Heidelberg, Germany

Printed by Libri Plureos GmbH
in Hamburg, Germany